Synthesis Lectures on Mathematics & Statistics

Series Editor

Steven G. Krantz, Department of Mathematics, Washington University, Saint Louis, MO, USA

This series includes titles in applied mathematics and statistics for cross-disciplinary STEM professionals, educators, researchers, and students. The series focuses on new and traditional techniques to develop mathematical knowledge and skills, an understanding of core mathematical reasoning, and the ability to utilize data in specific applications.

Arthur David Snider

Basics of Optimization Theory

 Springer

Arthur David Snider
University of South Florida
Tampa, FL, USA

ISSN 1938-1743   ISSN 1938-1751   (electronic)
Synthesis Lectures on Mathematics & Statistics
ISBN 978-3-031-29218-7   ISBN 978-3-031-29219-4   (eBook)
https://doi.org/10.1007/978-3-031-29219-4

This Springer imprint is published by the registered company Springer Nature Switzerland AG
The registered company address is: Gewerbestrasse 11, 6330 Cham, Switzerland

Preface

Optimization theory is the science of locating the extrema (maxima, minima) of a function. Since a minimum point for a function is a maximum point for its negative, the techniques for finding minima and maxima are essentially the same.

I believe that the easiest way to grasp the *basics* of any science is in terms of concrete visualizations—abstraction generalizations can come later. So whether we characterize a particular optimization algorithm as a minimization or maximization technique will be determined by the underlying model that we have chosen for motivating the algorithm's "game plan."

I hope my choices of models will grease the skids for you as you master Fibonacci search, the Simplex algorithm, steepest descent, conjugate gradient methods, and the Karush-Kuhn-Tucker-John conditions.

Tampa, USA
Arthur David Snider

Contents

Fibonacci Search

1

1.1   Unimodal Functions and Fibonacci Search

The first optimization technique we shall study in this book is called *Fibonacci search*. It is simple—no calculus, no matrices. It applies to a very broad class of functions. They don't have to be differentiable, or even continuous. And for this class of functions it is the *optimal* technique, providing the best estimate of the extreme point for a given number of function evaluations. It's also an amusing warmup for our subsequent chapters.

Truthfully, it is not used very much in applications, because usually the functions to be optimized *do* possess derivatives, which enables more efficient approaches.

We'll focus on finding the *minimum* point of an objective function $f(x)$. (Its maximum point is the minimum for $-f(x)$.) The class of functions suitable for Fibonacci search is exemplified in Figs. 1.1 and 1.2. The smooth example in Fig. 1.1 is simply a parabola $f(x) = ax^2 + bx + c$ $(a > 0)$ and its minimum value $\left(-b^2/4a + c\right)$ occurs at the location where the derivative is zero: $x_0 = -b/2a$. But the example in Fig. 1.2 isn't even continuous; its derivative—where it exists—is never zero. (Indeed, the actual minimum *value* of such a function could, in practice, never be determined by measurement unless we were told *exactly* where the minimum point x_0 was.)

However, both examples depict *unimodal* functions: such a function has, by definition, exactly one minimum point and is strictly monotonically decreasing on its left, and increasing on its right. Though we can't guarantee to approximate the minimal *value* of a unimodal function by measurement alone, we shall see that we can locate the *minimum point* x_0 as closely as required.

Suppose you were given a unimodal function on the unit interval $0 \leq x \leq 1$, and you could only afford to evaluate it at *one* point. Which point would you choose, and what could you conclude about the minimum from the evaluation? Take a minute and think about this.

© The Author(s), under exclusive license to Springer Nature Switzerland AG 2023
A. D. Snider, *Basics of Optimization Theory*, Synthesis Lectures on Mathematics & Statistics, https://doi.org/10.1007/978-3-031-29219-4_1

Fig. 1.1 Smooth unimodal function

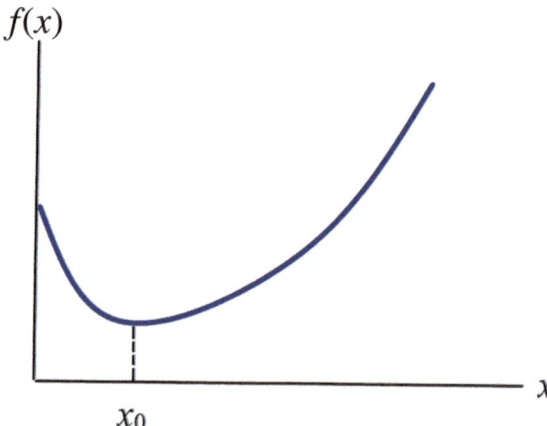

$f(x)$

x

x_0

Fig. 1.2 Discontinuous unimodal function

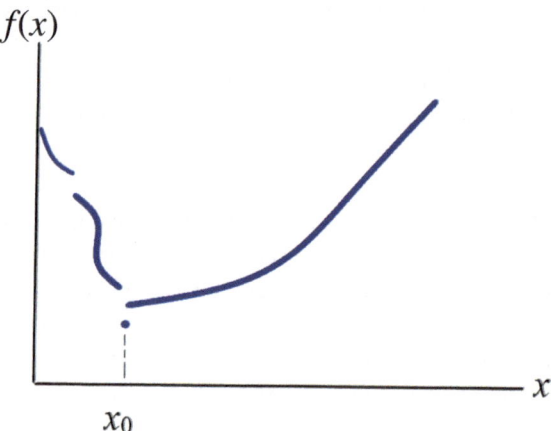

$f(x)$

x

x_0

Nothing? Right! A single evaluation tells you nothing about where the minimum lies—the interval of uncertainty remains the same: namely, [0, 1]. You're going to need at least two evaluations to learn anything about the minimum.

If the two evaluations give results such as depicted in Fig. 1.3, where $f(x_2)$ exceeds $f(x_1)$, the minimum *must* lie to the left of x_2. To see why, try to draw a graph of a unimodal function passing through these points, whose minimum lies to the right of x_2.

But the minimum lies to the right of x_2 in Fig. 1.4.

In general, if a unimodal function has $f(x_2) > f(x_1)$, the interval beside x_2 *opposite* to x_2 can be eliminated in the search for the minimum; see Fig. 1.5.

(If $f(x_2)$ happens to equal $f(x_1)$, we can eliminate *both* exterior intervals; but of course this is exceptional.)

Fig. 1.3 Minimum point $< x_2$

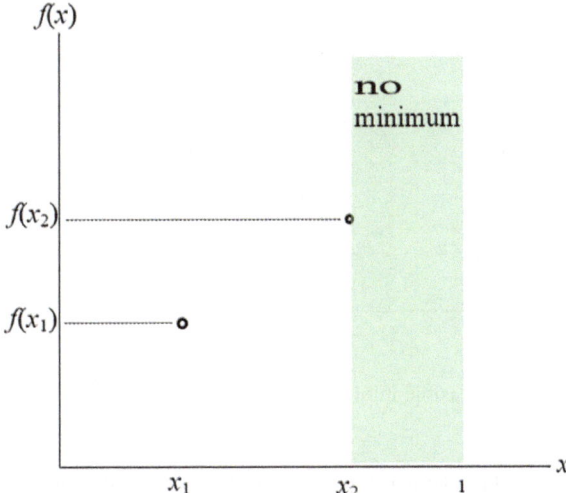

Fig. 1.4 Minimum point $> x_2$

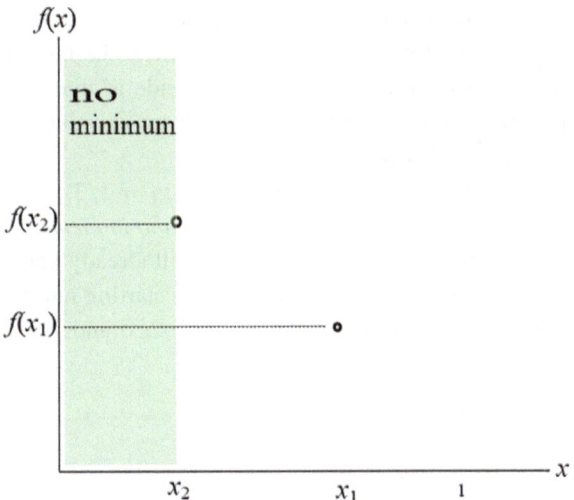

So, if you could afford *two* function evaluations in your search for the minimum point, where would you place them? Obviously x_1 and x_2 should be placed symmetrically. But the natural choice $x_1 = 1/3$, $x_2 = 2/3$ is *not* optimal.

The best choice for x_1 and x_2, ensuring the largest possible "disqualified" interval and the narrowest bracketing of the minimum point, is $x_1 = 0.5$, $x_2 = 0.5 \pm \varepsilon$, where ε is the length of the finest interval over which your measurement apparatus can distinguish the function values. (Don't worry too much about the ε details; they are carefully scrutinized in Avriel and Wilde (1966).)

We'll express this, somewhat imprecisely, as

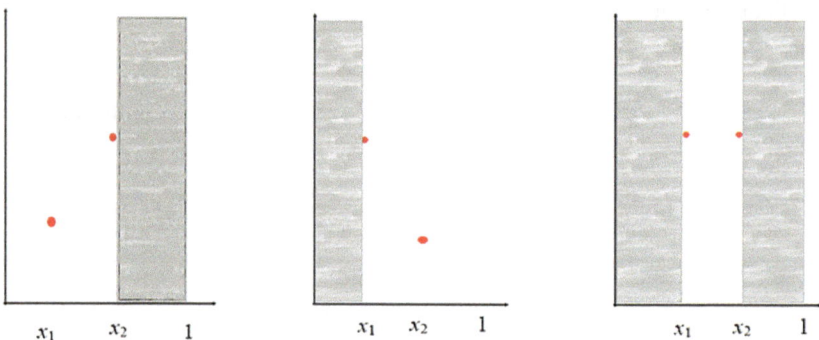

Fig. 1.5 Feasible minimum locations

The optimal placement of two function evaluations is side by side at the midpoint, reducing the interval of uncertainty for the minimum point by 50%. See Fig. 1.6.

Now if you could afford *three* function evaluations, where would you place them? A "greedy" approach would be to eliminate the largest subinterval at the outset by placing the first two measurements side by side at the "midpoint," as in Fig. 1.6. But then the third measurement, if it yields a still lower value for f, does not enable you to eliminate a subinterval! (Fig. 1.7.)

An alternative would be to choose $x_1 = 1/3$, $x_2 = 2/3$. This only eliminates 1/3 of the interval, *but it leaves you in a position to implement the optimal two-point placement with your third measurement*, since you will already know the midpoint value. In other words, if you eliminated the right third of the starting interval with the first pair of measurements, you are looking at the subinterval [0, 2/3] and you know the value at 1/3. So place your

Fig. 1.6 Optimal placement of
two measurements

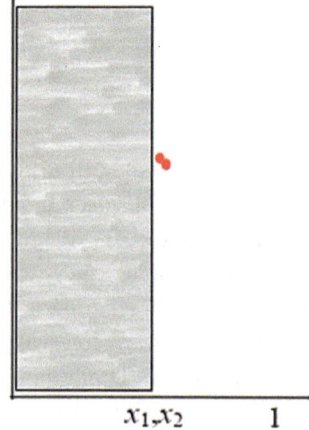

Fig. 1.7 Non-optimal placement of three measurements

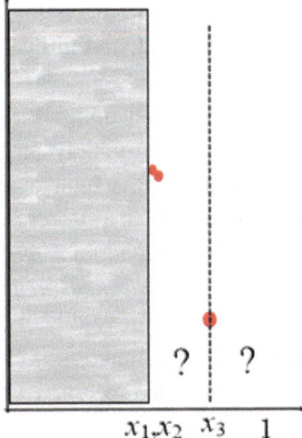

final measurement just to the right of 1/3, and reduce the resolution interval down to 1/3 (Fig. 1.8).

In fact in the next section we will prove that this is, indeed, the optimal three-point scheme. *We choose the first two evaluations for the three-point scheme so that the resulting elimination puts us in position to implement the optimal two-point scheme with the next measurement.*

Okay, how do we choose x_1 and x_2 for the four-point scheme, so as to put ourselves in position to implement the optimal three-point scheme? We need to choose x_1 and x_2 symmetrically so that after, say, we eliminate the interval $[x_2, 1]$, then x_1 is positioned at either 1/3 or 2/3 of *the remaining interval*. So: can you guess where x_1 and x_2 go?

Fig. 1.8 Optimal placement of three measurements

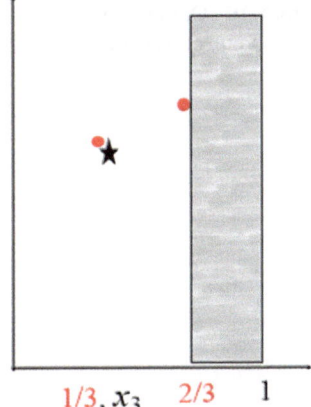

Did you come up with $x_1 = 1/4$, $x_2 = 3/4$? That's not bad. It eliminates 1/4 of the original interval, and ultimately locates the minimum point to within an interval of length 1/4; look at Fig. 1.9.

But that's not optimal! The choice $x_1 = 2/5$, $x_2 = 3/5$ eliminates 2/5 of the original interval, and ultimately locates the minimum point to within an interval of length 1/5; study Fig. 1.10, which depicts the result if $f(3/5) > f(2/5)$, and $f(2/5) > f(1/5)$. The minimum is located to a resolution of 1/5.

The best N-point scheme appears to be one where, after eliminating one subinterval with the first two measurements, we are in position to implement the best (N − 1)-point scheme. We'll prove this in the next section. Experiment with the best 5-point and 6-point schemes for a function $f(x)$ whose minimum point lies to the far right. You'll get the results tabulated in Fig. 1.11.

Fig. 1.9 Non-optimal placement of four measurements

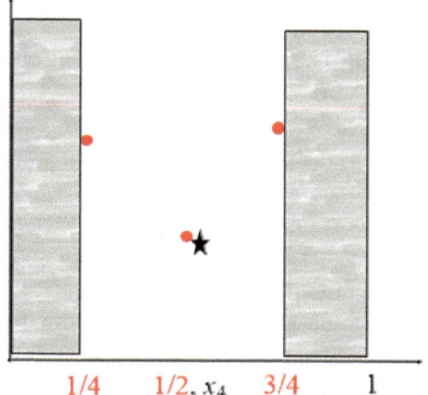

Fig. 1.10 Optimal four measurement placement

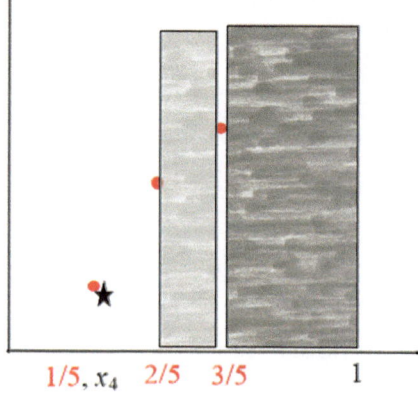

# of measurements   **resolution**

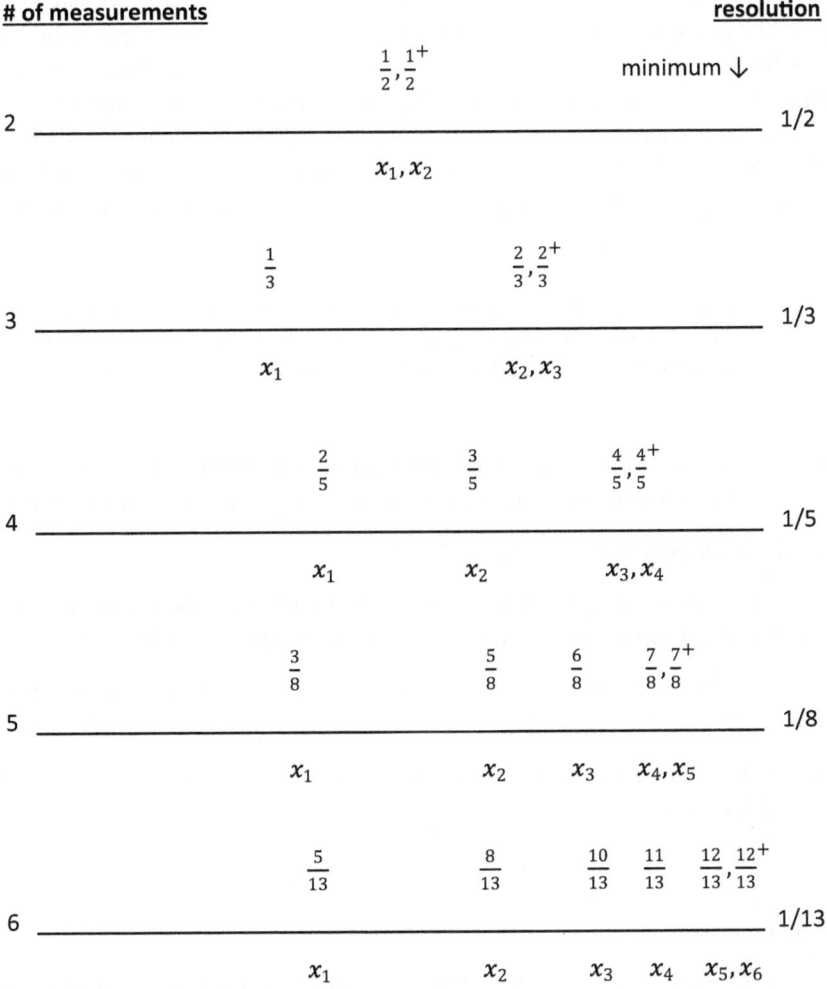

Fig. 1.11 Fibonacci measurement placements

The pattern is revealed in the reciprocals of the resolution lengths: 2, 3, 5, 8, 13, ... Each number is the sum of the two preceding numbers, and if we start off with 0, 1, 1 we get the *Fibonacci sequence*:

$F_0 = 0, F_1 = 1, F_2 = 1, F_3 = 2, F_4 = 3, F_5 = 5, F_6 = 8, F_7 = 13, F_8 = 21, F_9 = 34, F_{10} = 55, F_{11} = 89, F_{12} = 144, F_{13} = 233 \ldots$

$$\Rightarrow F_n = F_{n-2} + F_{n-1}$$

This sequence was studied in 1202 by the mathematician Fibonacci (Leonardo da Pisa), to model rabbit populations (Sigler 2003). It is well known in biological modeling, appearing in tree branching, arrangement of leaves on a stem, pineapple and artichoke fruitlets, ferns, pinecones, honeybee populations, and nautilus spirals. Devotees have founded the *Fibonacci Society*, uncovering such trivia as the fact that the glamorous 1940s movie actress Veronica Lake's bust measurement was F_9 (inches), and her waist was F_8 (https://www.howtallis.org/veronica-lake-height-weight-shoe-size). For further details about the Society visit their web site.

The Fibonacci search algorithm for finding the minimum of a unimodal function in the unit interval with N evaluations initiates with measurements at the symmetric points $x_1 = F_{N-1}/F_{N+1}$ and $x_2 = F_N/F_{N+1}$, and isolates the minimum point in an interval of size $1/F_{N+1}$.

Question 1. If you use Fibonacci search on a unimodal objective function on the unit interval, to what resolution can you locate the minimum point with 12 measurements?

Answer. The resolution will be $1/F_{13} = 1/233$.

Question 2. How many measurements will be required to find the minimum point of a unimodal objective function in the unit interval, to a resolution of 0.02?

Answer. $F_{10} = 55$ is the first Fibonacci number whose reciprocal is smaller than 0.02. Thus 9 measurements are required.

Question 3. Use the minimum number of evaluations to locate the minimum of the unimodal function

$$f(x) = -16x^4 + 24x^3 + 247x^2 - 384x + 144, 0 < x < 1,$$

to a resolution of 0.2.

Answer. The resolution is $0.2 = \frac{1}{5} = 1/F_5$, so 4 measurements are required. The first two measurements are at $F_3/F_5 = 2/5 = 0.4$ and $F_3/F_5 = 3/5 = 0.6$; since

$$f(0.4) = 31.0464 > f(0.6) = 5.6304,$$

we know the minimum lies in [0.4, 1]. We have a measurement at 3/5, and we take the next measurement at $4/5 = 0.8$. Since

$$f(0.6) = 5.6304 > f(0.8) = 0.6144,$$

we know the minimum lies in [3/5, 1], and we have a measurement at the midpoint 4/5 = 0.8. We'll take the last measurement at, say, 0.80001:

$$f(4/5) = 0.6144 < f(0.80001) = 0.6146,$$

and therefore $0.6 < x_{minimum} < 0.80001$.

1.2   Details (Optional Reading): Proof of the Optimality of Fibonacci Search

The use of the Fibonacci sequence for minimum search was formulated and demonstrated to be optimal by Kiefer (1953); the proof was refined and extended by Avriel and Wilde (1966). Herein we sketch the basic strategy behind their deliberations.

To show that the Fibonacci search algorithm is best possible, it seems natural to try to exploit the recurrence relation $F_N = F_{N-1} + F_{N-2}$. So we're going to compare the accuracies to which searches with N, $N-1$, and $N-2$ function evaluations locate the minimum of a unimodal function $f(x)$, on the unit interval $[0, 1]$.

We're going to employ a trick that is so slick it seems larcenous. First of all consider the graph of the unimodal function on the left in Fig. 1.12.

In the center of the figure we have compressed the graph horizontally by a factor of x_2 and tacked on a "tail" that does not affect the minimum but retains unimodality. We further compress it in the rightmost figure. (x_1 and x_2 will be specified shortly.)

Two important observations:

the minimum value of the function on the left is the same as the minimum values of the two compressed functions;

if we were to locate the minimum point for the function in the center of the figure to within an interval of size δ, we would know the minimum point for the function on the left to within an interval of size δ/x_2.

Now suppose the *best possible scheme* for finding the minimum of a particular unimodal function f on $[0, 1]$ using N evaluations calls for measurements at the points $x_1, x_2, x_3, ..., x_N$, and locates the minimum to within δ_N. By the considerations of the previous section, we know that the first two points x_1 and x_2 are independent of the values

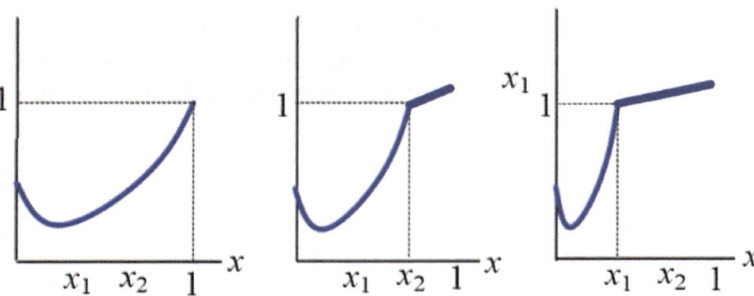

Fig. 1.12 Unimodal functions

of the function, but the choices for the remaining points $x_3, ..., x_N$ rely on the information gleaned from the previous measurements. We take $x_1 < x_2$; $x_1 + x_2 = 1$ by symmetry.

Now consider the three versions of the unimodal function depicted in Fig. 1.12.

(i) If we use the optimal scheme $\{x_1, x_2, x_3, ..., x_N\}$ on the original function on the left, we will locate the minimum point to within the optimal N-point measurement accuracy, which we have denoted by δ_N.

(ii) If we use the same scheme $\{x_1, x_2, x_3, ..., x_N\}$ to estimate the minimum point of the function in the center, we will locate *its* minimum to within δ_N, but that will only locate the minimum of the *original* function to within δ_N/x_2. Note however, that the measurement at x_2 is useless; so this is, in reality, an $(N-1)$-point scheme. Thus if δ_{N-1} denotes the *optimal* accuracy for an $(N-1)$-point scheme on the unit interval, then we must have $\delta_{N-1} \leq \delta_N/x_2$.

(iii) Now if we use $\{x_1, x_2, x_3, ..., x_N\}$ to locate the minimum point of the function on the right, we will (again) locate *its* minimum to within δ_N, thus locating the minimum of the original function to within δ_N/x_1. Note however, that the measurements at x_2 and x_1 can *both* be discarded, and this is, in reality, an $(N-2)$-point scheme. Thus we must have $\delta_{N-2} \leq \delta_N/x_1$.

Taking reciprocals, we have shown that the *optimal* location accuracies $\delta_N, \delta_{N-1}, \delta_{N-2}$ for such schemes are limited by

$$\frac{1}{\delta_{N-2}} + \frac{1}{\delta_{N-1}} \geq \frac{x_1}{\delta_N} + \frac{x_2}{\delta_N} = \frac{1}{\delta_N}, \text{ or } \delta_N \geq \frac{1}{\frac{1}{\delta_{N-2}} + \frac{1}{\delta_{N-1}}}. \tag{1.1}$$

(An upper limit on δ, of course, places a lower limit on $1/\delta$.)

Remember one evaluation provides no information about the minimum, so $\delta_1 = 1$; and we have seen that the optimal 2-evaluation search narrows the interval down to $\delta_2 = 1/2$. So the best resolution we can hope to get for 3 evaluations is δ_3, where $\frac{1}{\delta_3} \leq \frac{1}{1} + \frac{1}{1/2} = 1 + 2 = 3$ — but 3-point Fibonacci search provides resolution $1/3$, and that is optimal. The best possible resolution for a 4-point scheme is limited by $\frac{1}{\delta_4} \geq \frac{1}{1/2} + \frac{1}{1/3} = 2 + 3 = 5$ — again, the 4-point Fibonacci achieves the limiting value. Indeed, (1.1) shows that the reciprocals of the optimal search resolutions are, *at best*, limited by the Fibonacci recursion relation; and since the Fibonacci resolutions *fulfill* the relation, they are indeed optimal.

References

Avriel, M., Wilde, D.J.: Optimality proof for the symmetric Fibonacci search technique. Fibonacci Q. **4**, 265–269 (1966)

Fibonacci Society. http://www.maths.surrey.ac.uk/hosted-sites/R.Knott/Fibonacci/fibInArt.html. Accessed 27 Oct 2022

Kiefer, J.: Sequential minimax search for a maximum. Proc. Am. Math. Soc. **4**(3), 502–506 (1953)

Sigler, L.E.: Liber Abaci, Leonardo Pisano's Book of Calculation. Springer, New York (2003)

Linear Programming

2

2.1   An Example in Linear Programming

A dietician predicts that her Antarctic expedition team should consume 2300 oz of milk chocolate and 1200 oz of almonds during an upcoming 10-week expedition. Her outfitter can supply her with chocolate almond bars, each containing 1.08 oz of milk chocolate and 0.36 oz of almonds, for \$1.50 apiece; and he can supply bags of chocolate-covered almonds, each containing 2.75 oz of chocolate and 2.03 oz of almonds, for \$3.75 each. She can purchase either item in fractional quantities.

Question 1. How many chocolate bars and covered almonds should she purchase to meet the requirements—and how much will it cost?

Answer. A little calculation shows that 1,138.6 bars and 389.2 almonds exactly meets the requirements, and costs \$3,167.40.

> But would it interest you to know that she can save \$41 and meet—exceed, actually—the requirements, by purchasing 836.4 almonds and no bars, for a cost of only \$3,136.40?

This chapter explores *linear programming*, the technique for optimizing a linear cost function in the presence of linear *in*equality constraints.

2.2   The Two-Dimensional Linear Program

In Chap. 1 we saw how to search for the minimum point of a unimodal function without using sophisticated mathematics. Now we turn to the task of fathoming the extrema of a different type of function—the simple *linear function* $f(x) = ax + by$ (and its

© The Author(s), under exclusive license to Springer Nature Switzerland AG 2023   13
A. D. Snider, *Basics of Optimization Theory*, Synthesis Lectures on Mathematics
& Statistics, https://doi.org/10.1007/978-3-031-29219-4_2

higher dimensional generalizations), using basic matrix algebra. (And we continue to avoid calculus.)

Remember that the mathematical procedures for finding maxima and minima are essentially identical; we can maximize $f(x, y)$ by minimizing $-f(x, y)$. *To facilitate visualization, in this chapter we will speak in terms of finding* maxima.

First examine the level curves of $ax + by$. Of course, they are straight lines. Indeed, this observation is the source of the soubriquet "linear function". Figure 2.1 shows that the extrema are quite trivial; the minimum is $-\infty$ and the maximum is $+\infty$.

The maximization problem only acquires significance when we add *constraints* to the formulation, such as finding the maximum of $(x+y)$ inside the polygon drawn in Fig. 2.2.

From the figure we can see that the maximum is a little less than one.

You'll be happy to know that the simple configuration in Fig. 2.2 is prototypical of a large class of problems arising in practical applications. Conceptually, it is very easy to understand. Let's examine the specifics; then we'll formulate the general problem and outline the steps in its solution. The rest of the chapter will explore the details of the implementation of its solution—its *efficient* implementation, that is, because applications

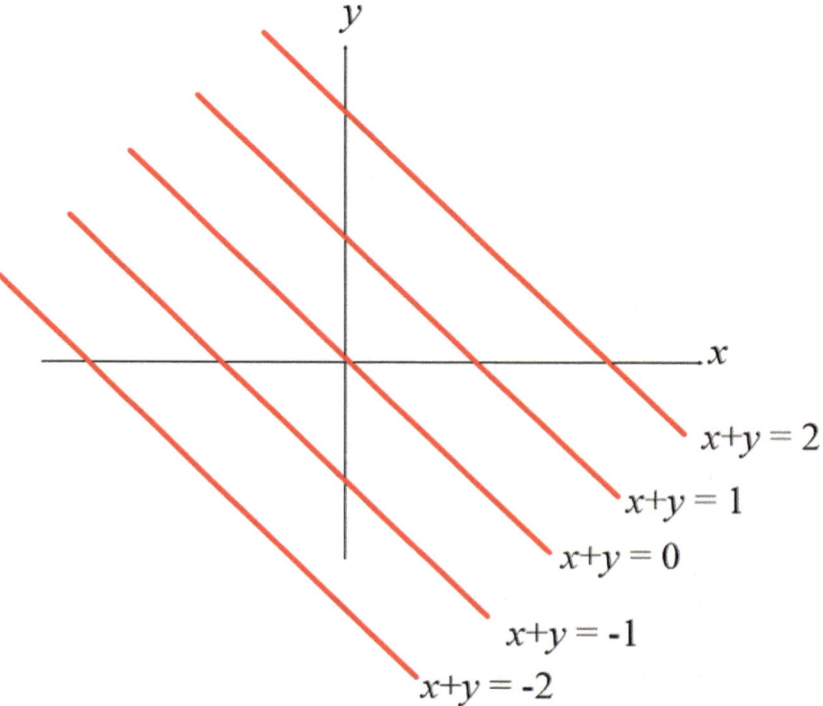

Fig. 2.1 Level curves of $x + y$

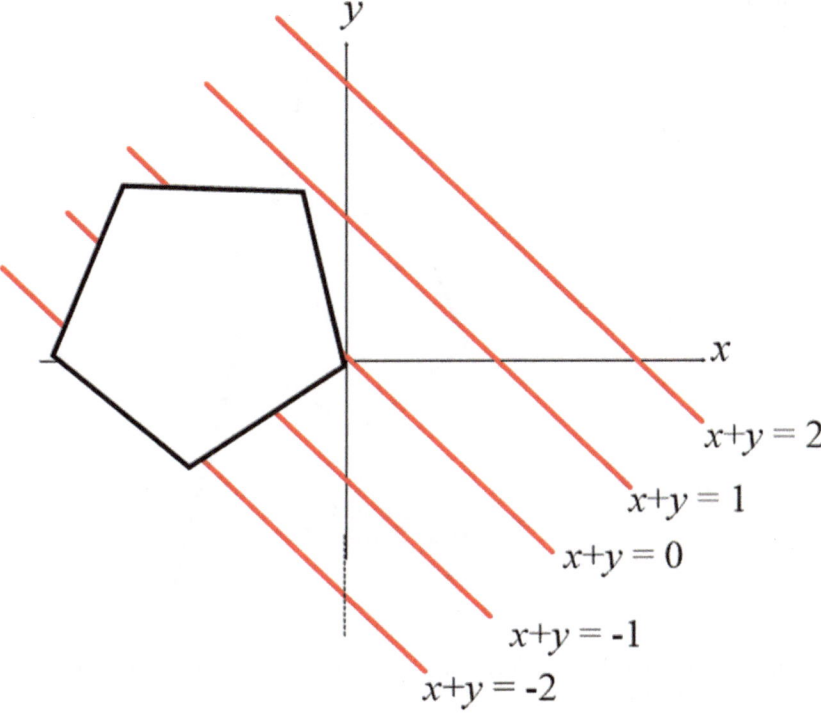

Fig. 2.2 Constrained maximization problem

of the algorithm arise in situations involving millions of unknowns, challenging computer capabilities.

The specifics are:

1. *The objective function to be maximized is linear:* $f(x, y) = ax + by$. For n variables (i.e., in n dimensions) the objective function takes the form $f(x_1, x_2, \cdots, x_N) = a_1 x_1 + a_2 x_2 + \cdots a_n x_N = \sum_{i=1}^{N} a_i x_i$.
2. *The variables are constrained to the interior of a region having the shape of a convex polygon.* We take some time to elaborate on this.

The boundary of a convex polygon is a collection of segments of straight lines. Figure 2.3 demonstrates the mathematical details for describing the relation between the interior and the straight line: if the equation of the straight line is $cx + dy = e$, then the "half-spaces" bounded by the line are described by the linear inequalities $cx + dy \leq e$ and $cx + dy \geq e$.[1]

---

[1] (We include the line itself in the half space).

Fig. 2.3 The half-spaces
defined by the constraint

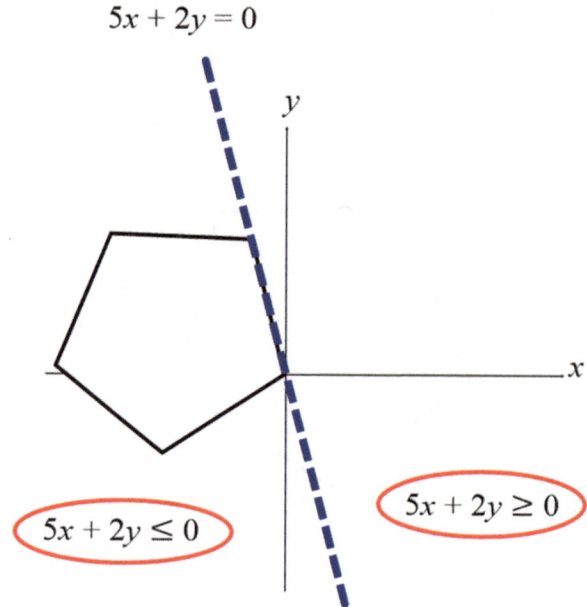

The form $cx + dy \le e$ is less restricted than it may seem. Several other constraint formulas can be recast this way:

$$cx + dy \ge e \Leftrightarrow -cx - dy \le -e;$$

$$cx + dy = e \Leftrightarrow cx + dy \le e \text{ and } x + dy \ge e;$$

$$e_1 \le cx + dy \le e_2 \Leftrightarrow cx + dy \le e_2 \text{ and } -cx - dy \le -e_1;$$

and since any (unrestricted or "*free*") number can be written as the difference of two nonnegative numbers,

$$x \text{ free } \Leftrightarrow x = x_1 - x_2, \; x_1 \ge 0 \text{ and } x_2 \ge 0.$$

For the sake of brevity we shall generally refer to the $cx + dy \le e$ form in this chapter. However the other forms may be simpler to work with in some situations.

The whole convex polygon is completely described by the linear inequalities corresponding to each of its boundary segments. In Fig. 2.4 the equations for the sides of the polygon are displayed.

Figure 2.5 displays the *crucial* data from Fig. 2.4, together with the level curves of the objective function $x + y$.

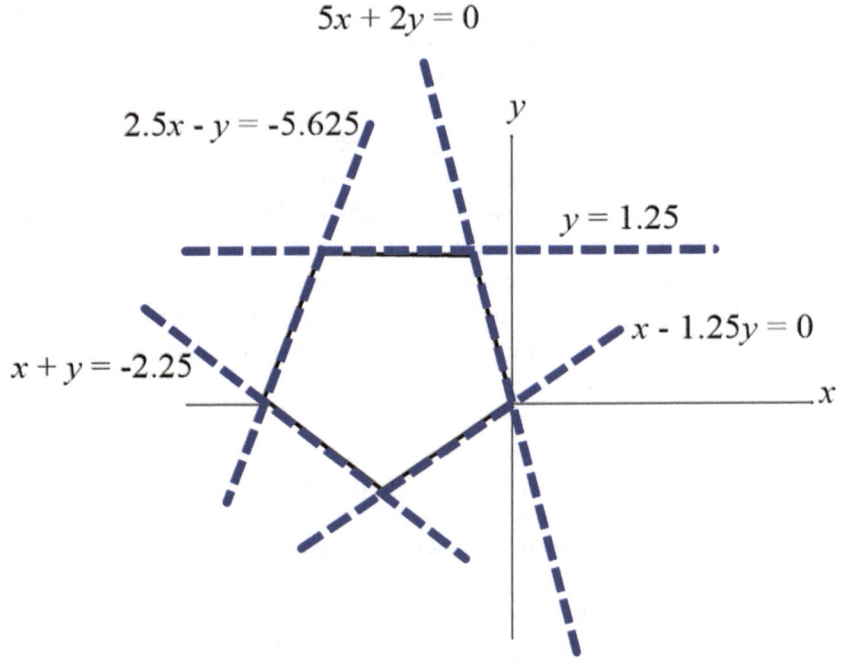

Fig. 2.4 Polygon formulas

Fig. 2.5 The optimum value is
achieved at a corner

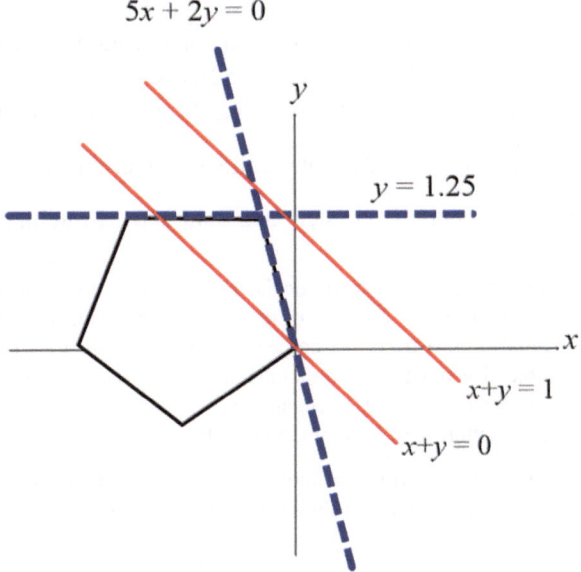

3. *The optimal value of the objective function is taken at a corner of the polygon.* In Fig. 2.5
   the maximum corner lies at the intersection of the lines $5x + 2y = 0$ and $y = 1.25$;
   that is, at $x = -0.5$, $y = 1.25$. The maximum value of $x + y$ is thus 0.75.

It is possible that the level curve for the optimal objective function could coincide with
one of the polygon's boundary segments. In Fig. 2.5, for instance, if the objection function
had been simply $f(x, y) = y$, the maximum (1.25) would occur all along the top edge.
This does not, however, contradict our statement; the maximum *does* occur at a corner—at
two corners, in fact, as well as along the whole boundary segment.

Our observations appear to make the solution procedure for linear programming problems quite simple: find all the corners, evaluate the objective function there, and select
the maximum.

The number of corners is clearly finite. In fact, since each corner is the intersection of two
boundary lines, if there are m inequality constraints then the number of *line pairs* is $\frac{m(m-1)}{2}$,
or $(5)(4)/2 = 10$ for the convex polygon in Fig. 2.4. The number of *corners* is fewer (only 5),
because some of the line pairs intersect outside the polygon.

Why, then, is linear programming a "science," meriting lucrative employment for specialists? There is a bogeyman hiding in that innocent-looking formula for the number of
corners that have to be tested. If there are n variables subject to m constraints, the number
of intersections to be fathomed is $\binom{m}{n}$, and this number can be huge if the number of
constraints is large. One of the first applications of linear programming involved blending
aviation gases (Charnes et al. 1952), and dealt with 22 unknowns, 22 constraints enforcing
their nonnegativity conditions, and 10 more "technological and policy restrictions." That's
$\binom{32}{22} = 64,512,240$ intersections(!). Current research addresses millions of unknowns
and constraints (Koberstein and Suhl 2007, Mittelmann 2022).

In 1947 George Dantzig devised the *Simplex* algorithm for solving linear programs (Dantzig
1951). Described as one of the top ten algorithms of the twentieth century (Nash 2000), the
algorithm restricts the corner search to intersections that increase the current value of the
objective function, and that actually lie on the polygon. We will study this remarkable algorithm shortly. Dantzig engagingly relates the story behind the Simplex algorithm, and linear
programming in general (Dantzig 1982).

In 1984 N. Karmarkar published—and patented—an improvement that allowed the solution
of still larger scale problems (Karmarkar 1984).

2.3   Review of Matrix Basics

As we observed at the end of the previous section, most important linear programming applications involve many variables and constraints. Thus it imperative that we employ matrix notation to keep track of the computations. Fortunately the requisite matrix tools are fairly elementary, and are covered in most linear algebra textbooks (Saff and Snider 2016). So you are probably already familiar with the content of the present section, and can proceed to Sect. 2.4, referring back here if you encounter some tricks that you have forgotten or some unfamiliar jargon. We shall forego formal proofs of the "theorems" and rely instead on presenting examples that will guide you to envisioning the appropriate generalizations.

We're not sticklers for mathematical notation. We'll indiscriminately represent a position vector as $\mathbf{R}$, $x\mathbf{i} + y\mathbf{j} + z\mathbf{k}$, $[x\ y\ z]$, $\begin{bmatrix} x \\ y \\ z \end{bmatrix}$, $[\mathbf{R} \rightarrow]$, $\begin{bmatrix} | \\ \mathbf{R} \\ \downarrow \end{bmatrix}$, and their n-dimensional versions. The context will obviate any ambiguity. Similarly the inner product $c_1 x + c_2 y + c_3 z$ may occur in the guise $\mathbf{c} \cdot \mathbf{R}$, $[c_1\ c_2\ c_3] \begin{bmatrix} x \\ y \\ z \end{bmatrix}$, $\mathbf{c}^T \mathbf{R}$, or $[c_1\ c_2\ c_3] \cdot [x\ y\ z]$ — regardless of how the two vectors have been represented. The identity matrix symbol $\mathbf{I}$ could denote $\mathbf{I}_{2\times 2}$, $\mathbf{I}_{3\times 3}$, or $\mathbf{I}_{n\times n}$ — again, depending on the context. Its columns are the vectors $\mathbf{e}_i$ with a "1" in its ith row and zeros elsewhere:

$$\mathbf{I}_{3\times 3} = \begin{bmatrix} 1 & 0 & 0 \\ 0 & 1 & 0 \\ 0 & 0 & 1 \end{bmatrix} = \begin{bmatrix} | & | & | \\ \mathbf{e}_1 & \mathbf{e}_2 & \mathbf{e}_3 \\ \downarrow & \downarrow & \downarrow \end{bmatrix}. \tag{2.1}$$

We do not burden $\mathbf{e}_i$ with further notation to indicate its number of rows; the context will make it clear.

A. Matrix Multiplication

The common interpretation of $\mathbf{Ax}$ is as a stack of inner products:

$$\begin{bmatrix} \mathbf{A}_{row\,1} \rightarrow \\ \mathbf{A}_{row\,2} \rightarrow \\ \vdots \\ \mathbf{A}_{row\,m} \rightarrow \end{bmatrix} \begin{bmatrix} | \\ | \\ \mathbf{x} \\ | \\ \downarrow \end{bmatrix} = \begin{bmatrix} \mathbf{A}_{row\,1} \cdot \mathbf{x} \\ \mathbf{A}_{row\,2} \cdot \mathbf{x} \\ \vdots \\ \mathbf{A}_{row\,m} \cdot \mathbf{x} \end{bmatrix} = \begin{bmatrix} b_1 \\ b_2 \\ \vdots \\ b_m \end{bmatrix}. \tag{2.2}$$

However it is equally valid to view Ax as a linear combination of the columns of A, with coefficients drawn from the vector x:

$$
\begin{bmatrix} | & | & & | \\ | & | & & | \\ A_{col\,1} & A_{col\,2} & \cdots & A_{col\,n} \\ | & | & & | \\ \downarrow & \downarrow & & \downarrow \end{bmatrix} \begin{bmatrix} | \\ | \\ x \\ | \\ \downarrow \end{bmatrix} = x_1 \begin{bmatrix} | \\ | \\ A_{col\,1} \\ | \\ \downarrow \end{bmatrix} + \cdots x_n \begin{bmatrix} | \\ | \\ A_{col\,n} \\ | \\ \downarrow \end{bmatrix} = \begin{bmatrix} | \\ | \\ b \\ | \\ \downarrow \end{bmatrix}. \tag{2.3}
$$

(Try it out.)

The focus of this section is the solution of a system of m linear equations in n unknowns,

$$
\left.\begin{array}{c} a_{11}x_1 + a_{12}x_2 + \cdots + a_{1n}x_n = b_1 \\ a_{21}x_1 + a_{22}x_2 + \cdots + a_{2n}x_n = b_2 \\ \vdots \\ a_{m1}x_1 + a_{m2}x_2 + \cdots + a_{mn}x_n = b_m \end{array}\right\}
$$

or

$$
\begin{bmatrix} a_{11} & a_{12} & \cdots & a_{1n} \\ a_{21} & a_{22} & \cdots & a_{2n} \\ & & \vdots & \\ a_{m1} & a_{m2} & \cdots & a_{mn} \end{bmatrix} \begin{bmatrix} x_1 \\ x_2 \\ \vdots \\ x_n \end{bmatrix} \begin{bmatrix} b_1 \\ b_2 \\ \vdots \\ b_m \end{bmatrix} \quad \text{or} \quad Ax = b. \tag{2.4}
$$

In these contexts A is the *coefficient matrix*. A more compact way of representing the system utilizes the *augmented matrix*

$$
\begin{bmatrix} a_{11} & a_{12} & \cdots & a_{1n} & | & b_1 \\ a_{21} & a_{22} & \cdots & a_{2n} & | & b_2 \\ & & & \vdots & | & \vdots \\ a_{m1} & a_{m2} & \cdots & a_{mn} & | & b_m \end{bmatrix}. \tag{2.5}
$$

Note that the a_{ij}'s are *coefficients* of the variables while the b_j's are stand-alone *constants*.

*Pre*multiplication of a matrix B by a matrix A takes place column-by-column:

$$
\begin{bmatrix} & A & \end{bmatrix} \begin{bmatrix} a & b & c \\ d & e & f \\ g & h & i \end{bmatrix} = \begin{bmatrix} A\begin{pmatrix} a \\ d \\ g \end{pmatrix} & A\begin{pmatrix} b \\ e \\ h \end{pmatrix} & A\begin{pmatrix} c \\ f \\ i \end{pmatrix} \end{bmatrix}. \tag{2.6}
$$

(*Post*multiplication takes place row-by-row.)

B. **Basic Solutions**

Frequently in our applications the number of unknowns (n) in a linear system exceeds the number of equations (m), and the system has an infinite number of solutions.[2] It often happens in the linear systems arising in linear programming calculations that the coefficient matrix contains all the columns of the m-by-m identity matrix, as well as other columns which we shall designate as "*cluttered*".[3] When this occurs, one particular solution is very obvious:

Question 2. In 15 seconds or less, find a solution to the system

$$
\begin{bmatrix}
1 & 0 & 4 & 0 & 5 & 0 & -1 \\
0 & 0 & 6 & 0 & 2 & 1 & 0 \\
0 & 0 & 2 & 1 & 1 & 0 & 1 \\
0 & 1 & 2 & 0 & 3 & 0 & 3
\end{bmatrix}
\begin{bmatrix}
x_1 \\ x_2 \\ x_3 \\ x_4 \\ x_5 \\ x_6 \\ x_7
\end{bmatrix}
=
\begin{bmatrix}
4 \\ 3 \\ 2 \\ 1
\end{bmatrix} .
\tag{2.7}
$$

Answer. This is too easy. Interpret the left hand side of (2.7) in the format (2.3). Do you spot the identity matrix in columns 1, 2, 4, and 6? You can express the right hand side $[4321]^T$ using only these columns by choosing the coefficients $x_1 = 4$, $x_2 = 1$, $x_4 = 2$, $x_6 = 3$, and all the other x's equal to 0.

Similarly, a solution of

$$
\begin{bmatrix}
0 & 3 & 1 & 0 \\
1 & 2 & 0 & 0 \\
0 & 4 & 0 & 1
\end{bmatrix}
\begin{bmatrix}
x_1 \\ x_2 \\ x_3 \\ x_4
\end{bmatrix}
=
\begin{bmatrix}
2 \\ -4 \\ 5
\end{bmatrix} ,
\tag{2.8}
$$

using columns 1, 3, and 4, is given by $x_1 = -4$, $x_3 = 2$, $x_4 = 5$, and $x_2 = 0$.

Of course both (2.7) and (2.8) have an infinite number of solutions, but we contend that the ones we listed are the most obvious.

Whenever each of the columns of the identity matrix appears (in any order) in the coefficient matrix for a linear system, there is always an obvious solution with the variables corresponding

---

[2] (Unless the system is simply inconsistent).

[3] This terminology is useful because it is picturesque. Be aware, though, that even a minus sign will "clutter" an identity column; for our purposes, the column $(-e_1)$ is regarded as *cluttered*.

to these columns equal to the corresponding data on the right-hand side, and the remaining variables equal to zero. Such a solution is called a basic solution.[4] The variables corresponding to the identity columns are called the *basic variables.*

("Basic solution" refers to the fact that from the perspective of linear algebra, the identity columns constitute the customary *basis* for the matrix's column space; the basic solution uses only these columns to express the right-hand side of the linear system. Without dwelling on these details, we shall employ the terminology "basic solution" throughout this chapter, to be consistent with other texts.)

C. **Pivoting**

Gauss elimination refers to an algorithm for the solution of linear systems that employs only three *elementary row operations*; in fact each of them can be expressed as a premultiplication by a matrix:

(i) multiply a row of a matrix by a nonzero constant. This can be accomplished through premultiplication by a matrix of the form

$$\begin{bmatrix} 1 & 0 & 0 \\ 0 & a & 0 \\ 0 & 0 & 1 \end{bmatrix} . \quad \text{(Try it out).} \tag{2.9}$$

(ii) add a multiple of one row to another. Again, premultiplication by the matrix

$$\begin{bmatrix} 1 & 0 & 0 \\ 0 & 1 & a \\ 0 & 0 & 1 \end{bmatrix} \tag{2.10}$$

adds a times the third row to the second. (Try it out.)

(iii) rearrange the order of the rows. This is equivalent to premultiplication by a *permutation matrix* of the form

---

[4] If any identity column occurs twice, there will be two basic solutions. For example, $\begin{bmatrix} 1 & 0 & 1 & \vdots & a \\ 0 & 1 & 0 & \vdots & b \end{bmatrix}$

has the basic solutions $x_1 = a, x_2 = b, x_3 = 0$ and $x_1 = 0, x_2 = b, x_3 = a$.

$$\begin{bmatrix} 0 & 1 & 0 \\ 0 & 0 & 1 \\ 1 & 0 & 0 \end{bmatrix}. \tag{2.11}$$

((Sigh) Try it out.) Such premultiplication of the augmented matrix of a linear system preserves its solutions, because each of the elementary row operations is invertible. (Try to guess the matrix inverses of (2.9)–(2.11) from the corresponding operations, without any calculations.)

The *pivoting* maneuver applies a combination of the elementary row operations to replace a cluttered column in a matrix by a column of the identity. As an example, let's replace the fifth column of the matrix below with e_3:

$$\begin{bmatrix} 1 & 0 & 0 & a & b & c \\ 0 & 1 & 0 & d & e & f \\ 0 & 0 & 1 & g & h & i \end{bmatrix} \Longrightarrow \begin{bmatrix} \times & \times & \times & \times & 0 & \times \\ \times & \times & \times & \times & 0 & \times \\ \times & \times & \times & \times & 1 & \times \end{bmatrix}. \tag{2.12}$$

(For reasons to be seen, we intentionally selected a matrix containing all the columns of the identity.) First we'll drive "b" to 0 by subtracting b/h times the third row from the first row. Following the logic of (2.10), this is equivalent to premultiplying (2.12) by

$$\begin{bmatrix} 1 & 0 & -b/h \\ 0 & 1 & 0 \\ 0 & 0 & 1 \end{bmatrix}. \tag{2.13}$$

Next we annihilate the "e" by subtracting e/h times the third row from the second row—or premultiplying by

$$\begin{bmatrix} 1 & 0 & 0 \\ 0 & 1 & -e/h \\ 0 & 0 & 1 \end{bmatrix}. \tag{2.14}$$

Finally we divide the third row by h, to replace the "h" by 1; premultiply the matrix by

$$\begin{bmatrix} 1 & 0 & 0 \\ 0 & 1 & 0 \\ 0 & 0 & 1/h \end{bmatrix}. \tag{2.15}$$

In the jargon of matrix computations, the (3, 5) entry "h" is called the *pivot entry* (or simply "pivot"). Pivoting on the (i, j) entry replaces the jth column (*pivot column*) by

e_i; the ith row (*pivot row*) is simply rescaled. Pivoting, as described, is not possible if the pivot entry is zero.[5]

It is instructive to see what the *composite* pivoting matrix premultiplier, and the final matrix that results, look like. Being careful to write the premultipliers in the proper order and applying the *associative law* of matrix multiplication we have

$$
\begin{bmatrix}
1 & 0 & 0 & a & b & c \\
0 & 1 & 0 & d & e & f \\
0 & 0 & 1 & g & h & i
\end{bmatrix} \Longrightarrow
$$

$$
\underbrace{\begin{bmatrix} 1 & 0 & 0 \\ 0 & 1 & 0 \\ 0 & 0 & 1/h \end{bmatrix} \begin{bmatrix} 1 & 0 & 0 \\ 0 & 1 & -e/h \\ 0 & 0 & 1 \end{bmatrix} \begin{bmatrix} 1 & 0 & -b/h \\ 0 & 1 & 0 \\ 0 & 0 & 1 \end{bmatrix}}_{\Big\downarrow} \begin{bmatrix} 1 & 0 & 0 & a & b & c \\ 0 & 1 & 0 & d & e & f \\ 0 & 0 & 1 & g & h & i \end{bmatrix}
$$

$$\text{(2.16)}$$

$$
= \begin{bmatrix} 1 & 0 & -b/h \\ 0 & 1 & -e/h \\ 0 & 0 & 1/h \end{bmatrix} \times \begin{bmatrix} 1 & 0 & 0 & a & b & c \\ 0 & 1 & 0 & d & e & f \\ 0 & 0 & 1 & g & h & i \end{bmatrix}
$$

$$
= \begin{bmatrix}
1 & 0 & -b/h & a - gb/h & 0 & c - ib/h \\
0 & 1 & -e/h & d - ge/h & 0 & f - ie/h \\
0 & 0 & 1/h & g/h & 1 & i/h
\end{bmatrix}.
$$

Several important observations can be drawn from (2.16):

(a) **Pivot Matrix Structure**

The composite premultiplier matrix that achieves the pivoting operation (red in display (2.16)) has one cluttered column, and the remainder are identity columns. If we had pivoted on any other nonzero entry in the third *row*, the columns of the pivoting matrix would have the same general form as (2.16) — $[e_1, e_2, cluttered]$. If we had pivoted on an entry in the *second* row, the columns would take the form $[e_1, cluttered, e_3]$. If we pivoted on both in succession, the columns would look like $[e_1, cluttered, cluttered]$. (Try it out.)

(b) **Pivoting an Identity Column**

The pivot ("h") is the (3, 5) entry of the matrix. The pivot column (the fifth column) is transformed into the identity column e_3 (as intended); any copies of e_3 in the original

---

[5] And it can be inaccurate if h is small.

matrix are transformed into cluttered columns; but all the other identity columns are unaffected.[6]

(c) **Equation Content Immunity**

If the matrix in (2.16) happens to be the augmented matrix for a linear system of equations, then pivoting on its (3, 5) entry is equivalent to solving the *third* equation for the *fifth* variable

$$
\begin{bmatrix} 1 & 0 & 0 & a & b & | & c \\ 0 & 1 & 0 & d & e & | & f \\ 0 & 0 & 1 & g & h & | & i \end{bmatrix} \implies x_5 = -\frac{1}{h}x_3 - \frac{g}{h}x_4 + \frac{i}{h}, \tag{2.17}
$$

and substituting for it in the other equations for example, the first equation becomes

$$
x_1 + 0x_2 + 0x_3 + ax_4 + b\left[-\frac{1}{h}x_3 - \frac{g}{h}x_4 + \frac{i}{h}\right] = c
$$

or

$$
x_1 + 0x_2 - b\frac{1}{h}x_3 + \left(a - b\frac{g}{h}\right)x_4 = c - b\frac{i}{h}
$$

(which agrees with the first row of (2.16)).

In short: after a pivoting operation is performed on an augmented matrix, each equation expresses the same condition as it did before pivoting, but the condition is articulated in terms of different combinations of the variables. Or, in the case of the pivot row, the equation is simply rescaled.

(d) **Column Location Immunity**

If the positions of the columns of a matrix are reordered prior to pivoting, the result will be the same as if the pivoting were performed on the original matrix prior to the reordering:

$$
\begin{bmatrix} 1 & 0 & 0 & a & b & c \\ 0 & 1 & 0 & d & e & f \\ 0 & 0 & 1 & g & h & i \end{bmatrix} \underset{on\,"h"}{\overset{pivot}{\Longrightarrow}} \begin{bmatrix} 1 & 0 & -b/h & a-gb/h & 0 & c-ib/h \\ 0 & 1 & -e/h & d-ge/h & 0 & f-ie/h \\ 0 & 0 & 1/h & g/h & 1 & i/h \end{bmatrix}
$$

---

[6] A silly exception: if the original pivot column were simply a multiple of e_3, other multiples of e_3 are merely rescaled.

[*reorder the columns, then pivot*]

$$\begin{bmatrix} 1 & 0 & 0 & c & a & b \\ 0 & 1 & 0 & f & d & e \\ 0 & 0 & 1 & i & g & h \end{bmatrix} \overset{pivot}{\underset{on\,"h"}{\Longrightarrow}} \begin{bmatrix} 1 & 0 & -b/h & c-ib/h & a-gb/h & 0 \\ 0 & 1 & -e/h & f-ie/h & d-ge/h & 0 \\ 0 & 0 & 1/h & i/h & g/h & 1 \end{bmatrix}.$$

(e) Row Location Immunity

Similarly to the above, if the positions of the rows of a matrix are reordered prior to pivoting, the result will be the same as if the pivoting were performed on the same datum before the reordering:

$$\begin{bmatrix} 1 & 0 & 0 & a & b & c \\ 0 & 1 & 0 & d & e & f \\ 0 & 0 & 1 & g & h & i \end{bmatrix} \overset{pivot}{\underset{on\,"h"}{\Longrightarrow}} \begin{bmatrix} 1 & 0 & -b/h & a-gb/h & 0 & 0 \\ 0 & 1 & -e/h & d-ge/h & 0 & f-ie/h \\ 0 & 0 & 1/h & g/h & 1 & i/h \end{bmatrix}$$

[*reorder the rows, then pivot*]

$$\begin{bmatrix} 0 & 0 & 1 & g & h & i \\ 1 & 0 & 0 & a & b & c \\ 0 & 1 & 0 & d & e & f \end{bmatrix} \overset{pivot}{\underset{on\,"h"}{\Longrightarrow}} \begin{bmatrix} 0 & 0 & 1/h & g/h & 1 & i/h \\ 1 & 0 & -b/h & a-gb/h & 0 & 0 \\ 0 & 1 & -e/h & d-ge/h & 0 & f-ie/h \end{bmatrix}$$

Pivoting can also be used to find *basic solutions* when the number of unknowns in a system exceeds the number of equations. In display (2.18) we pivot on the $(1, 1)$ and the $(2, 3)$ entries:

$$\begin{bmatrix} 1 & 2 & 2 & 2 & | & 1 \\ 3 & 3 & 4 & 2 & | & 0 \end{bmatrix} \overset{pivot}{\Longrightarrow} \begin{bmatrix} 1 & -1 & 0 & -2 & | & -2 \\ 0 & 1.5 & 1 & 2 & | & 1.5 \end{bmatrix}; \tag{2.18}$$

$$\begin{bmatrix} x_1 \\ x_2 \\ x_3 \\ x_4 \end{bmatrix}_{basic} = \begin{bmatrix} -2 \\ 0 \\ 1.5 \\ 0 \end{bmatrix}.$$

D. Row Reduction

Suppose we have a linear system $Ax = b$ containing the same number of equations as unknowns, so that its augmented matrix

$$\begin{bmatrix} a_{11} & a_{12} & \cdots & a_{1m} & | & b_1 \\ a_{21} & a_{22} & \cdots & a_{2m} & | & b_2 \\ & & \vdots & & | & \vdots \\ a_{m1} & a_{m2} & \cdots & a_{mm} & | & b_m \end{bmatrix} \equiv [A|b] \quad (2.5 \text{ repeated})$$

contains a *square* nonsingular coefficient submatrix A. If we can pivot on each of the main diagonal entries in succession, A's columns are replaced by identity columns $e_1, e_2, \ldots, e_m$[7]:

$$\begin{bmatrix} 1 & 0 & \cdots & 0 & | & ? \\ 0 & 1 & \cdots & 0 & | & ? \\ & & & \vdots & | & \vdots \\ 0 & 0 & \cdots & 1 & | & ? \end{bmatrix} \equiv [I\,|\,(?)]\,.^{11} \tag{2.19}$$

What happens to the final column? Answer: pivoting preserves the solution and (2.19) states $Ix = [final\ column]$; but $Ix = x$, so *the final column is the solution* $A^{-1}b$. This procedure—pivoting down the diagonal of the coefficient submatrix (and switching rows when necessary) to obtain the solution—is known as *Gauss-Jordan elimination* or *the row-reduction algorithm.*

Display (2.20) below exhibits two examples of the workings of the row-reduction algorithm:

$$\begin{bmatrix} 1\ 2\ |\ 1 \\ 3\ 4\ |\ 0 \end{bmatrix} \overset{Gauss-Jordan}{\Longrightarrow} \begin{bmatrix} 1\ 0\ |\ -2 \\ 0\ 1\ |\ 1.5 \end{bmatrix}; \quad \begin{bmatrix} x_1 \\ x_2 \end{bmatrix} = \begin{bmatrix} -2 \\ 1.5 \end{bmatrix}, \tag{2.20}$$

$$\begin{bmatrix} 0\ 2\ |\ 1 \\ 2\ 4\ |\ 0 \end{bmatrix} \Rightarrow \begin{bmatrix} 2\ 4\ |\ 0 \\ 0\ 2\ |\ 1 \end{bmatrix} \overset{G-J}{\Longrightarrow} \begin{bmatrix} 1\ 0\ |\ -1 \\ 0\ 1\ |\ 0.5 \end{bmatrix}; \quad \begin{bmatrix} x_1 \\ x_2 \end{bmatrix} = \begin{bmatrix} -1 \\ 0.5 \end{bmatrix}.$$

As a spinoff of these observations, we have a slick way to compute a matrix's inverse. Thanks to the column-by-column interpretation of matrix multiplication (2.6), the formula $AA^{-1} = I$ identifies the columns of A^{-1} as the solutions to the equations $Ax_i = e_i$:

$$I_{m \times m} = [e_1 \cdots e_m], \tag{2.21}$$

---

[7] If a zero pops up on the diagonal, we simply scan the pivot column for a nonzero entry, and exchange rows. (If *all* the entries in our pivot column are zero, the matrix is singular).

$$AA^{-1} = A\left[A_{col\,1}^{-1} \cdots A_{col\,m}^{-1} \right] = \left[A\underbrace{\left(\begin{matrix} | \\ A_{col\,1}^{-1} \\ \downarrow \end{matrix} \right)}_{=e_1} \cdots A\underbrace{\left(\begin{matrix} | \\ A_{col\,m}^{-1} \\ \downarrow \end{matrix} \right)}_{=e_m} \right].$$

Thus the columns of A^{-1} (i.e., the solution) will appear on the right if we apply Gauss-Jordan elimination to the compound system with the augmented form below on the left:

$$[A\,|\,e_1 e_2 \cdots e_m] \equiv [A\,|\,I] \overset{Gauss-Jordan}{\Rightarrow} \left[I\,|\,A^{-1} \right]. \tag{2.22}$$

For example, pivoting down the main diagonal of the matrix below on the left produces the display

$$\begin{bmatrix} 1\,2\,|\,1\,0 \\ 3\,4\,|\,0\,1 \end{bmatrix} \overset{Gauss-Jordan}{\Rightarrow} \begin{bmatrix} 1\,0\,|\,-2 & 1 \\ 0\,1\,|\,1.5 & -0.5 \end{bmatrix}, \tag{2.23}$$

from which we deduce the (easily verified) formula

$$\begin{bmatrix} 1 & 2 \\ 3 & 4 \end{bmatrix}^{-1} = \begin{bmatrix} -2 & 1 \\ 1.5 & -0.5 \end{bmatrix}. \tag{2.24}$$

The computation tools discussed in this section will be exploited when we examine the matrix formulations of the Simplex algorithm later in this chapter.

2.4   The Diamond Analogy and the Simplex Method

Figures 2.1–2.5 in Sect. 2.1 address a two-dimensional linear program. They mask some of the subtleties of the Simplex code, so in this section we describe a three-dimensional model that will enable us to understand the workings of the general algorithm.

Suppose you have a diamond resting on the table, and you are looking for the highest point on the diamond; *the objective function to be maximized is the vertical (z) coordinate.* We shall see that this task is typical of all linear programming problems. First of all we need to characterize the geometry of a diamond.[8]

A "diamond in the rough" is a shapeless carbon stone that takes on its beautiful sparkle when it is properly cut. The diamond's crystallography results in its having four families of parallel planes along which it is structurally weak. It can easily be split, or cleaved,

---

[8] Some of our figures and narrative are drawn from the web sites. http://en.wikipedia.org/wiki/Diamond_cut and http://www.amazon.com/The-Rise-Fall-Diamonds-ebook/dp/B0050CC56A/ref=sr_1_12?s=digital-text&ie=UTF8&qid=1350499352&sr=1-12&keywords=edward+jay+epstein.

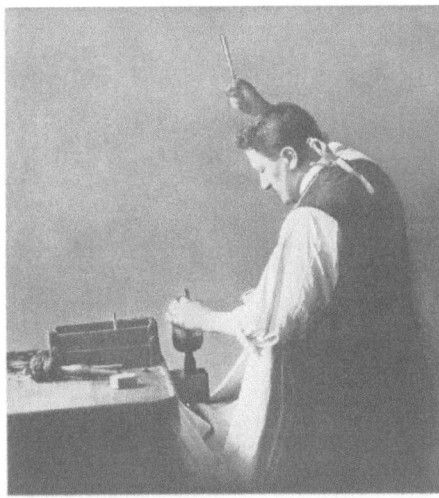

Fig. 2.6 **a** "Joseph Asscher using a hammer to make the first split of the Cullinan diamond" (https://commons.wikimedia.org/wiki/File:Joseph_Asscher_splitting_the_Cullinan_diamond.jpg) by unknown author (1908) (https://archive.org/details/TheCullinan/page/n15/mode/2up) is in the public domain. **b** Cleaved diamond (https://archive.org/details/TheCullinan/page/n11/mode/2up) by unknown author (1908) is in the public domain

using a fine chisel, along these planes (Fig. 2.6). There is an illustrative video of the process at http://www.langantiques.com/university/index.php/A_History_Of_Diamond_Cutting.

Each of these weakly bonded planes is parallel to one of the faces of an octahedron. Ancient jewelers produced ornamental diamonds by cleaving the stone along these faces, creating the *point cut* illustrated in Fig. 2.7 and spawning the familiar idiom "diamond shape."

As the technology of diamond cutting progressed, gemologists were able to produce the shapes in Fig. 2.8 by pressing the rough diamond against a *scaif*, a spinning polishing wheel sprinkled with diamond dust. They could thus grind other planar faces, or *facets*, into the stone. Once every point on the rough surface has been cleaved or ground down to a plane face, the dazzling jewel can reflect or refract light from all directions.

We are going to confine our attention to gem shapes created by cleaving or grinding. To be sure, the invention of the diamond saw enabled the manufacture of more exotic gem shapes like the heart in Fig. 2.8, where some of the plane cuts do not go all the way through the body. But we'll assume that *through every point on our diamond surface there is a plane such that the entire diamond lies on one side of the plane.* These are called *supporting planes*, each defining a *half-space* containing the diamond; indeed, the diamond is the mathematical intersection of all the half-spaces of its supporting planes.

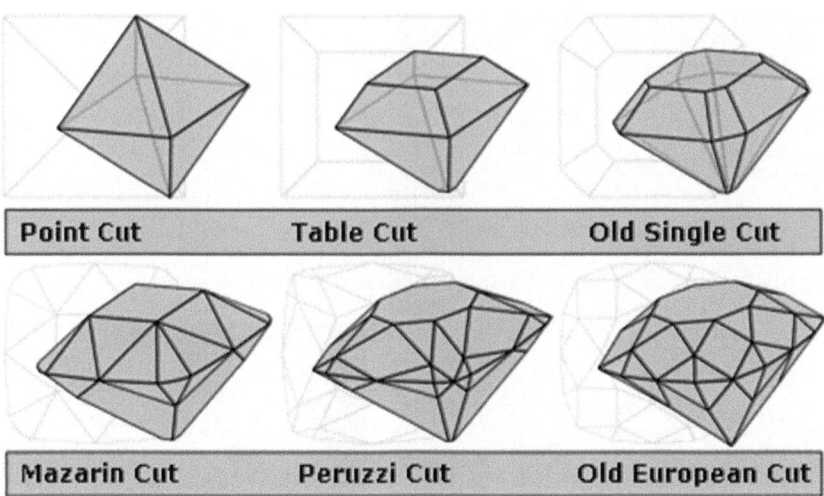

Fig. 2.7 "A diagram of old diamond cuts showing the evolution from the most primitive (point cut) to the most advanced pre-Tolkowsky cut (old European)" by Juergen Schoner (July 2004), with English translation by Gregory Phillips (https://commons.wikimedia.org/wiki/File:Dia_Hist.png) is licensed under CC BY-SA 3.0 Unported license (https://creativecommons.org/licenses/by-sa/3.0/leg alcode)

Fig. 2.8 "Fancy Cuts" (https:// commons.wikimedia.org/wiki/ File:Fancy_cut_diamonds.jpg) by Paul Noillimrev (Sept. 2012) (own work) is licensed under CC BY-SA 3.0 Unported license (https://creativecomm ons.org/licenses/by-sa/3.0/leg alcode)

Such a diamond is a polyhedron, and it is **convex**: if two points lie inside the diamond, then the entire line segment joining them also lies inside.

Now the strategy for the Simplex algorithm for finding the highest corner on such a diamond proceeds as follows. Every corner of the diamond is the junction of three or more edges. Starting from any corner, we pick one of the edges that heads "uphill." We proceed along this uphill edge until we reach the next corner. Then we repeat. If none of the edges emanating from our corner heads uphill, we stop, because we are at the highest point.

That's the essence of the Simplex method. We'll address a few troublesome complications later; for now we want to keep it simple. Let's formulate these steps using the language of vector algebra. First we devise a mathematical description of a diamond created by cleaving and grinding, i.e. a convex polyhedron.

In Fig. 2.9, n denotes a vector of unit length normal to a supporting plane; the plane is perpendicular to the page, and n points *away from* the diamond. The distance from the plane to the origin is d. The tip (x, y, z) of the generic vector $R = x\mathbf{i} + y\mathbf{j} + z\mathbf{k}$ then lies on the "diamond-side" of the plane if $n \cdot R \leq d$.[9] The diamond is thus described by the simultaneous solutions to the inequalities (see Fig. 2.10)[10]

$$
\begin{aligned}
half-space\,\#1: \quad & n_1 \cdot R \leq d_1 \quad or \quad && a_1 x + b_1 y + c_1 z \leq d_1 \\
half-space\,\#2: \quad & n_2 \cdot R \leq d_2 \quad or \quad && a_2 x + b_2 y + c_2 z \leq d_2 \\
half-space\,\#3: \quad & n_3 \cdot R \leq d_3 \quad or \quad && a_3 x + b_3 y + c_3 z \leq d_3 \\
& \quad\quad\quad\vdots \\
half-space\,\#M: \quad & n_M \cdot R \leq d_M \;\; or \quad && a_M x + b_M y + c_M z \leq d_M,
\end{aligned}
\tag{2.25}
$$

$$
or \quad
\begin{bmatrix}
a_1 & b_1 & c_1 \\
a_2 & b_2 & c_2 \\
a_3 & b_3 & c_3 \\
& \vdots & \\
a_M & b_M & c_M
\end{bmatrix}
\begin{bmatrix} x \\ y \\ z \end{bmatrix}
\leq
\begin{bmatrix} d_1 \\ d_2 \\ d_3 \\ \vdots \\ d_M \end{bmatrix}.
$$

A *corner* on the diamond is characterized by the intersection of at least three supporting planes. But three planes can intersect in several ways (or not at all), as depicted in Fig. 2.11. For them to define a corner they must have one and only one point in common.

---

[9] Recall $a \cdot b$ equals (length of a) times (length of b) times (cosine of the angle between a and b). We have drawn Fig. 2.9 with $d > 0$, to facilitate visualization. The reader is invited to sketch the layout for $n \cdot R \leq d$ with $d < 0$, and confirm that n still points away from the diamond. Therefore d can be positive or negative in the inequality $n \cdot R \leq d$; nonetheless n always points *away* from the diamond.

[10] We take the liberty of interchanging matrix and geometric vector notation freely. Thus for example $R, x\mathbf{i} + y\mathbf{j} + z\mathbf{k}, \begin{bmatrix} x & y & z \end{bmatrix}, and \begin{bmatrix} x & y & z \end{bmatrix}^T$ all refer to the same quantity.

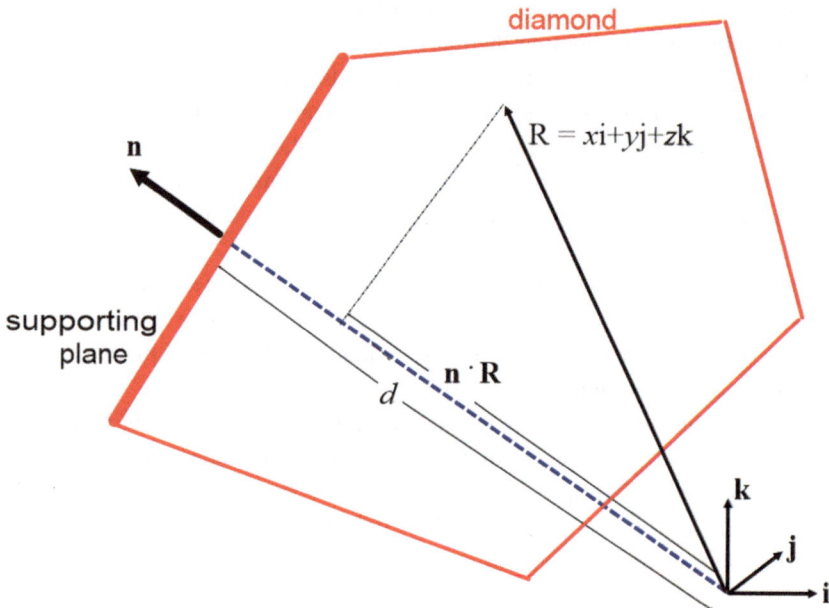

Fig. 2.9 Vector description of a convex polyhedron

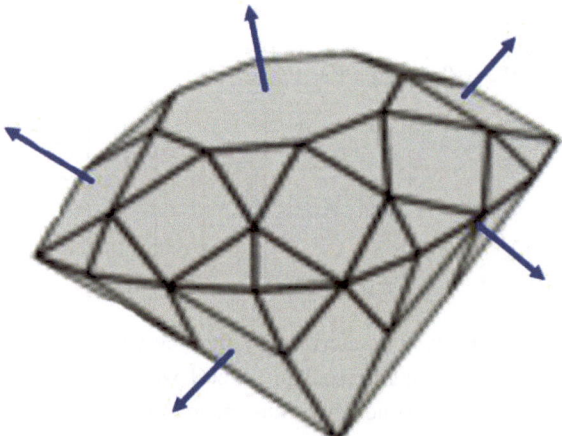

Fig. 2.10 "A diagram of old diamond cuts showing the evolution from the most primitive (point cut) to the most advanced pre-Tolkowsky cut (old European)" by Juergen Schoner (July 2004), with English translation by Gregory Phillips (https://commons.wikimedia.org/wiki/File:Dia_Hist. png), used under CC BY-SA 3.0 Unported license (https://creativecommons.org/licenses/by-sa/3.0/ legalcode). Cropped and embellished from original

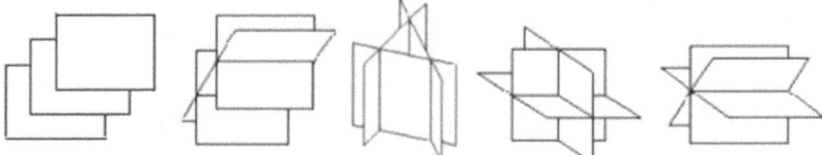

Fig. 2.11 Three intersecting planes

Question 3. What are the conditions on the equations for three planes that guarantee they intersect in one point?

Answer. There are many equivalent ways of expressing the condition. If the equations of three planes are

$$n_1 \cdot R = d_1, \quad n_2 \cdot R = d_2, \quad n_3 \cdot R = d_3, \tag{2.26}$$

or
$$\begin{bmatrix} n_1 \rightarrow \\ n_2 \rightarrow \\ n_3 \rightarrow \end{bmatrix} \begin{bmatrix} | \\ R \\ \downarrow \end{bmatrix} \equiv \underbrace{\begin{bmatrix} a_1\ b_1\ c_1 \\ a_2\ b_2\ c_2 \\ a_3\ b_3\ c_3 \end{bmatrix}}_{[N]} \begin{bmatrix} x \\ y \\ z \end{bmatrix} \equiv [N] \begin{bmatrix} x \\ y \\ z \end{bmatrix} = \begin{bmatrix} d_1 \\ d_2 \\ d_3 \end{bmatrix},$$

then they intersect in a single point—i.e. they define a corner—if the determinant of the coefficient matrix N is not zero; equivalently, if the coefficient matrix is nonsingular, or if its rows or columns are linearly independent, or if the normal vectors are not coplanar.

Any triplet of *equalities* from the set of inequalities (2.25) specifies a single point if these conditions are met. And the point is a corner—that is, it lies on the diamond—if the remaining *in*equalities (2.25) are satisfied.

> Of course some distinct triplets may define the same corner; indeed, many of the corners in the table-cut diamond of Fig. 2.7 lie in more than three supporting planes. This is one of the troublesome complications that we'll ignore until Sect. 2.11.

We move from one corner to another by traveling along an edge. An edge emanates from a corner if it is a line of intersection of two of the supporting planes defining the corner. So if the corner is specified by the three equations in (2.26), the edge that lies in, say, the first two planes must be perpendicular to those planes' normals, n_1 and n_2. Similarly the edge lying in the second and third planes is perpendicular to n_2 and n_3.

Question 4. How does one construct a perpendicular to a given pair of normals?

Answer. This is too easy; take the cross product. But that trick won't work in higher dimensions. The Gram-Schmidt algorithm can be browbeaten into providing an answer, but for our purposes the best construction is to use the inverse of the coefficient matrix

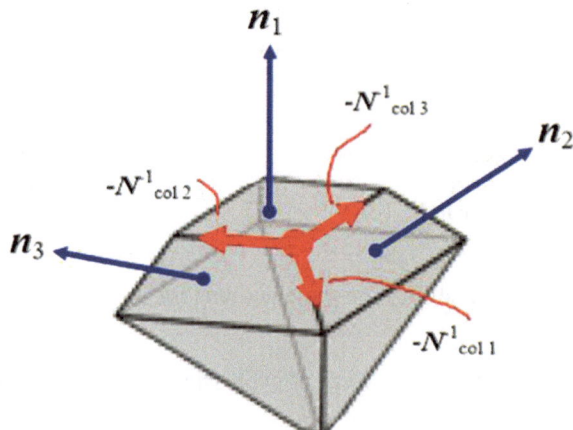

Fig. 2.12 Supporting plane normal and edge vectors. ("A diagram of old diamond cuts showing the evolution from the most primitive (point cut) to the most advanced pre-Tolkowsky cut (old European)" by Juergen Schoner (July 2004), with English translation by Gregory Phillips (https://commons.wikimedia.org/wiki/File:Dia_Hist.png), used under CC BY-SA 3.0 Unported license (https://creativecommons.org/licenses/by-sa/3.0/legalcode). Cropped and embellished from original.)

N in (2.26). Study the pattern of zeros in the final (identity) matrix in the matrix-inverse formula

$$
\begin{bmatrix} n_1 \rightarrow \\ n_2 \rightarrow \\ n_3 \rightarrow \end{bmatrix} \begin{bmatrix} n_1 \rightarrow \\ n_2 \rightarrow \\ n_3 \rightarrow \end{bmatrix}^{-1} \equiv \begin{bmatrix} n_1 \rightarrow \\ n_2 \rightarrow \\ n_3 \rightarrow \end{bmatrix} [N]^{-1} = \begin{bmatrix} n_1 \rightarrow \\ n_2 \rightarrow \\ n_3 \rightarrow \end{bmatrix} \underbrace{\begin{bmatrix} | & | & | \\ | & | & | \\ \downarrow & \downarrow & \downarrow \end{bmatrix}}_{[N^{-1}]} = \begin{bmatrix} 1 & 0 & 0 \\ 0 & 1 & 0 \\ 0 & 0 & 1 \end{bmatrix}.
$$

$$(2.27)$$

The first column of the inverse is perpendicular to ... *which* ... row vectors?

Right; second and third! And its second column is perpendicular to n_1 and n_3, and so on. So *the columns* $\begin{bmatrix} N^{-1}_{col1} & N^{-1}_{col2} & N^{-1}_{col3} \end{bmatrix}$ *of the matrix* N^{-1} *provide the directions of the edges emanating from a corner.* See Fig. 2.12 (whose minus signs will be explained).

Now we need to know if N^{-1}'s column vectors—when attached to our corner—point *away* from the diamond or along its edge. We have specified that the *normals* n_1, n_2, n_3, all point away from the diamond (Figs. 2.9, 2.12), and the presence of the *positive* diagonals in display (2.27) indicates that each column vector in the inverse makes an acute angle with the corresponding normal (interpret the row-column products as dot products[11]). For example, the inverse's first column makes an acute angle with n_1; it

---

[11] We repeat: $a \cdot b$ equals (length of a) times (length of b) times (cosine of the angle between a and b).

points *away* from the diamond. So to travel *along the diamond's edges*, we must proceed *in the opposite directions* of the columns of N^{-1}, as indicated in Fig. 2.12.

Which of the three edges should we choose? Remember that our goal is to find the *highest* corner, that is, the one with the largest z-coordinate. Now it is clear from the convex geometry of the diamond that unless we have already reached the highest point (and we stop!), there will be at least one edge along which the z-coordinate increases. To find one, we look among our three edge-direction vectors—the negatives of the columns of N^{-1} — and choose one with a positive z-coordinate. That is, we choose one (any one) that makes a positive dot product with $k = \begin{bmatrix} 0 & 0 & 1 \end{bmatrix}$:

$$[k \rightarrow] \ \ times \ (-) \begin{bmatrix} n_1 \rightarrow \\ n_2 \rightarrow \\ n_3 \rightarrow \end{bmatrix}^{-1} \equiv \underbrace{[0\ 0\ 1]}_{k}(-1) \underbrace{\begin{bmatrix} | & | & | \\ | & | & | \\ \downarrow & \downarrow & \downarrow \end{bmatrix}}_{[N^{-1}]} = \begin{bmatrix} z_1 \\ z_2 \\ z_3 \end{bmatrix}; \qquad (2.28)$$

pick a positive z. For example, let us suppose z_2 is positive, so we proceed in the opposite direction of the second column of the matrix inverse; we move from R to $R + \delta R$, where δR is a positive multiple of $-N^{-1}_{col2}$:

$$\delta R = -N^{-1}_{col\ 2} \ times \ w \ (w > 0). \qquad (2.29)$$

Question 5. How do we determine w? How far should we go in this direction?

Answer. We encounter the next corner in the diamond when we bang into another supporting plane (Fig. 2.12)—so we simply increase w from zero until one of the remaining *in*equality constraints is on the verge of being violated.

The constraint for supporting plane #i reads $n_i \cdot R \le d_i$, allowing a degree of "slack"

$$s_i \equiv d_i - n_i \cdot R.$$

We can increase w in (2.29) so long as $n_i \cdot (R + \delta R)$ does not exceed d_i (that is, the slack s_i stays nonnegative) for *any* of the supporting planes with $4 \le i \le M$. (Of course, $s_i = 0$ for $i = 1, 2, 3$.)

We shall employ the nomenclature *slack variables* for the numbers s_i, and *coordinate variables* for x, y, and z.

To get an explicit expression for δR, we write (recall (2.29))

$$n_i \cdot \delta R \equiv n_i \cdot \left(-N^{-1}_{col\ 2}\right) w \le d_i - n_i \cdot R = s_i. \qquad (2.30)$$

If we define the scalar t_i by

$$t_i \equiv -\boldsymbol{n}_i \cdot \boldsymbol{N}_{col\,2}^{-1}, \ i = 4, 5, \ldots, M, \tag{2.31}$$

then (2.30) says we must maintain $t_i w \leq s_i$ for every i in order to stay on the diamond. Since $w > 0$ we have no problem with the values of i where $t_i \leq 0$, and we can increase w up to the lowest value of the ratio s_i/t_i for the positive t_i.

Question 6. What if none of t_i are positive?

Answer. In such a case there is no limit as to how far we can proceed along the chosen edge—we'll never encounter another corner. The diamond is infinite (unbounded) in this direction. The maximum z is infinity and the optimization search is complete. We'll document an illustration of this in Example 2.3, Sect. 2.9.

Question 7. What if, instead of seeking the *highest* point, we seek the most remote point in the direction of the vector $\boldsymbol{c}$?

Answer. In this case our objective is to maximize $\boldsymbol{c} \cdot \boldsymbol{R}$ instead of $z = \boldsymbol{k} \cdot \boldsymbol{R}$. So we replace $\boldsymbol{k}$ by $\boldsymbol{c}$ in Eq. (2.28),

$$-[\boldsymbol{c} \rightarrow] \begin{bmatrix} \boldsymbol{n}_1 \rightarrow \\ \boldsymbol{n}_2 \rightarrow \\ \boldsymbol{n}_3 \rightarrow \end{bmatrix}^{-1} = -[\boldsymbol{c} \rightarrow] \begin{bmatrix} \boldsymbol{N}_{col\,1}^{-1} & \boldsymbol{N}_{col\,2}^{-1} & \boldsymbol{N}_{col\,3}^{-1} \end{bmatrix}, \tag{2.32}$$

select a positive entry (say, the second), and proceed in the direction (2.29) as before.

Question 8. What if no entry in formula (2.28) is positive, but some entry is 0?

Answer. If, say, z_2 in (2.28) is zero and the other entries are negative, then although we have reached a corner where $\boldsymbol{k} \cdot \boldsymbol{R}$ is maximum, we can proceed along the direction $-\boldsymbol{N}_{col\,2}^{-1}$ without changing its value. That is, the maximum of the objective function is achieved by *all* points along this edge; the edge is horizontal. If we calculate $\delta \boldsymbol{R}$ by the previous formulas, we will reach the other corner along this edge. (Unless, as in Question 6, this horizontal edge extends to infinity.)

> That's the logic of the Simplex method. To facilitate expressing the algorithm with a matrix formulation in the next section, we summarize the equations here. For simplicity we assume that there are exactly three supporting planes passing through each corner. Workarounds for the exceptional situations will be addressed in later sections.

Summary: To maximize $\boldsymbol{c} \cdot \boldsymbol{R}$ subject to $\boldsymbol{n}_i \cdot \boldsymbol{R} \leq d_i, i = 1, \cdots, M$:

1. Pick a *corner* $\boldsymbol{R}$ of the diamond defined by the inequalities and renumber the inequalities so that the normals of the supporting planes passing through $\boldsymbol{R}$ are $\boldsymbol{n}_1, \boldsymbol{n}_2$, and $\boldsymbol{n}_3$.

Question 9. How do we pick the first corner?

Answer. We've been rather glib about this point; it's more subtle than may appear. We can't simply pick *any* three of the inequalities (with independent normals), because we don't know that the intersection point will satisfy the other inequalities. (Sometimes we shall use the terminology *legitimate corner* to emphasize that the intersection point satisfies all of the inequalities; the nomenclature *feasible solution* is also widely used. Until feasibility is verified, the intersection is simply a *corner candidate*.) The selection of the first corner is known as "Phase One" of the Simplex algorithm, and we'll defer discussion until Sect. 2.10. Meanwhile we'll choose examples where there is an obvious corner from which we can launch the search.

Pick any column index K (1, 2, or 3) giving rise to a positive entry in the vector

$$-[c \to] \begin{bmatrix} n_1 \to \\ n_2 \to \\ n_3 \to \end{bmatrix}^{-1} \equiv -[c_1 \, c_2 \, c_3] \Big[N^{-1}_{col\,1} \, N^{-1}_{col\,2} \, N^{-1}_{col\,3} \Big] = -c \cdot N^{-1}. \qquad (2.33)$$

If no entry is positive, STOP. The current corner is optimal. (As Question 8 explained, there may be other optimal points as well.)[12]

For this value of K, let e_k be the column of the 3-by-3 identity matrix with all zeros except for a 1 in its Kth entry:

$$I_{3\times3} = \begin{bmatrix} 1\,0\,0 \\ 0\,1\,0 \\ 0\,0\,1 \end{bmatrix}; \quad e_1 = \begin{bmatrix} 1 \\ 0 \\ 0 \end{bmatrix}, \, e_2 = \begin{bmatrix} 0 \\ 1 \\ 0 \end{bmatrix}, \, e_3 = \begin{bmatrix} 0 \\ 0 \\ 1 \end{bmatrix}.$$

2. Tabulate the numbers $t_i^{(K)}$ for each supporting plane not passing through the corner:

$$t_i^{(K)} \equiv -[n_i \to] \begin{bmatrix} n_1 \to \\ n_2 \to \\ n_2 \to \end{bmatrix}^{-1} \begin{bmatrix} | \\ e_K \\ \downarrow \end{bmatrix} = -n_i \cdot N^{-1}_{col\,K}, \quad i = 4, 5, 6, \cdots, M. \qquad (2.34)$$

If none of the $t_i^{(K)}$ are positive, STOP. The diamond is infinite and $c \cdot R$ is unbounded.

---

[12] As Question 8 explained, there may be other "equally optimal" points. It is possible that although no entry in (2.33) is positive, one may be zero. A quick review of the derivations in this section shows that in such a case the objective function stays constant along the corresponding edge of the diamond. Conducting the s/t competition for this edge reveals another corner where the objective is maximal; the entire edge consists of optimum points. *Or*, if none of the t are positive under these circumstances, the diamond is unbounded and the maximal objective is attained along this infinite edge.

3. Tabulate the slacks s_i afforded by the supporting planes not passing through this corner:

$$s_4 = d_4 - \boldsymbol{n}_4 \cdot \boldsymbol{R}$$
$$s_5 = d_5 - \boldsymbol{n}_5 \cdot \boldsymbol{R}$$
$$\vdots$$
$$s_M = d_M - \boldsymbol{n}_M \cdot \boldsymbol{R} \qquad (2.35)$$

4. Among the positive $t_i^{(K)}$, find the minimum of the ratios $s_i/t_i^{(K)}$ for the supporting planes not passing through the corner:

$$w = \min_i s_i/t_i^{(K)}, \quad 4 \le i \le M$$
$$\equiv s_I/t_I^{(K)}. \qquad (2.36)$$

5. Start over, searching from the corner at

$$\boldsymbol{R} + \delta\boldsymbol{R} = \boldsymbol{R} - w \begin{bmatrix} \boldsymbol{n}_1 \to \\ \boldsymbol{n}_2 \to \\ \boldsymbol{n}_3 \to \end{bmatrix}^{-1} \begin{bmatrix} | \\ \boldsymbol{e}_K \\ \downarrow \end{bmatrix} \equiv \boldsymbol{R} - \left(s_I/t_I^{(K)}\right) N_{col\,K}^{-1}. \qquad (2.37)$$

This corner passes through the supporting planes with normals from the set $\{\boldsymbol{n}_1, \boldsymbol{n}_2, \boldsymbol{n}_3\}$ with $\boldsymbol{n}_K$ replaced by $\boldsymbol{n}_I$. The slack s_K ceases to be zero; instead s_I becomes zero.

Renumber the constraints so that the normals passing through $\boldsymbol{R} + \delta\boldsymbol{R}$ are called $\{\boldsymbol{n}_1, \boldsymbol{n}_2, \boldsymbol{n}_3\}$, and repeat these steps....

For brevity, throughout this chapter we shall refer to the selection of the minimum of the $s_i/t_i^{(K)}$ ratios (2.36) as the *s/t* **competition**, and $s_I, t_I^{(K)}$ as the **competition winners**.

2.5   Simple Generalizations

To make the exposition easier to follow, we imposed some restrictions on the linear program explained in Sect. 2.3. Some of these can be easily generalized.

1. Presuming the normal vector $\boldsymbol{n}$ in the constraint $\boldsymbol{n} \cdot \boldsymbol{R} \le d$ to be a unit vector merely simplifies the interpretations in Fig. 2.9. This condition is not necessary for the Simplex formulation.

Question 10. Why not?

Answer. The answer is elementary but boring. If the "n_j" in the constraint $n_j \cdot R + s_j = d_j$ has length $|n_j| \neq 1$, the honest thing to do would be to replace $n_j, s_j,$ and d_j by $n_j/|n_j|, s_j/|n_j|,$ and $d_j/|n_j|$ throughout Sect. 2.4. If you do this, you will see that the (positive) factor $|n_j|$ can be canceled in practically every equation; the others (2.28), (2.29), (2.32), (2.33), (2.36) simply provide ($\pm$) signs or (s/t) ratio winners and are unaffected by the factor.

2. Since we are exploiting the diamond analogy in this introductory exposition, we shall continue to confine ourselves to 2 or 3 dimensions.[13] The extension to higher dimensions will become obvious when we present the *matrix tableau* implementation of the Simplex algorithm.
3. We have stated, without proof, a number of intuitively obvious properties concerning convex sets. Formal proofs can be found in the linear programming literature.

And of course we still owe you the explanation of how to find a starting corner and how to proceed when too many supporting planes pass through a corner.

Sections 2.6–2.8 are devoted to an exposition of the matrix implementation of Dantzig's Simplex algorithm. We must admit that these sections are not for everyone, although we have tried to motivate the steps. Readers with no taste for programming details may draw comfort in knowing that the diamond-analogy logic, with all the refinements, has been reliably coded in commercial software; and they may "safely" skip ahead to Sect. 2.12. For the rest, we of the optimization fraternity bid a warm welcome. We presume you are familiar with the tools of Sect. 2.3, and we recommend a pot of strong coffee.

2.6 The Basic Simplex Algorithm

We are going to describe Dantzig's Simplex algorithm, a matrix implementation of our diamond-corner search procedure. The goal of the formulation presented in this section is to highlight the role that each computation plays in finding the maximum objective function. In this section we take a leisurely pace, ignoring shortcuts that may occur to you. Rest assured they will be incorporated in Sects. 2.7–2.9, where a sleek, streamlined version is derived.[14]

We presume that the linear program has been formulated as

$$\text{maximize } c \cdot R \tag{2.38}$$

---

[13] The notation is going to be formidable enough, without attempting complete generality.

[14] Experts will note many roundabout deviations from standard expositions of this theory.

subject to the inequality constraints

$$
\begin{aligned}
\boldsymbol{n}_1 \cdot \boldsymbol{R} \le d_1 &\quad or \quad a_1 x + b_1 y + c_1 z \le d_1 \\
\boldsymbol{n}_2 \cdot \boldsymbol{R} \le d_2 &\quad or \quad a_2 x + b_2 y + c_2 z \le d_2 \\
\boldsymbol{n}_3 \cdot \boldsymbol{R} \le d_3 &\quad or \quad a_3 x + b_3 y + c_3 z \le d_3 \\
&\quad \vdots \\
\boldsymbol{n}_M \cdot \boldsymbol{R} \le d_M &\quad or \quad a_M x + b_M y + c_M z \le d_M,
\end{aligned}
\tag{2.39}
$$

that we are starting from a corner satisfying the *equality* forms of the first three constraints, and that none of the corners lie in more than three supporting planes.

The inventors of the Simplex algorithm discovered that all of the calculations in Sect. 2.4 can be achieved by the steps used in Gauss-Jordan elimination (Sect. 2.3). Now we can't indiscriminately do these manipulations with *in*equalities. (For example, multiplying an inequality by a negative constant reverses the direction of the inequality.) So a first step in implementing the matrix formulation is to reformulate all of the inequalities as equalities.

We accomplish this by using the *slack variables* s_i mentioned above Eq. (2.43) of Sect. 2.4;

$$
if \ \boldsymbol{n}_i \cdot \boldsymbol{R} \le d_i, then \ \boldsymbol{n}_i \cdot \boldsymbol{R} + s_i = d_i \ with \ s_i \ge 0.
$$

Thus if the ith supporting plane actually *contains* the corner at $\boldsymbol{R}$, then

(i) the ith inequality is already an equality; and
(ii) the slack variable s_i is zero.

The constraint is said to be *active* in this case. If $s_i > 0$ it is *inactive*. When we move along an edge from one corner to another, we deactivate one constraint and activate another. (And remember we have assumed that no more than 3 constraints will be active at any corner.)

So the matrix formulation for M inequality constraints defining a convex polyhedron in 3 dimensions (i.e., a diamond) becomes a linear system of M *equations* in $M+3$ unknowns:

$$
\begin{aligned}
\boldsymbol{n}_1 \cdot \boldsymbol{R} + s_1 &= d_1 \\
\boldsymbol{n}_2 \cdot \boldsymbol{R} + s_2 &= d_2 \\
\boldsymbol{n}_3 \cdot \boldsymbol{R} + s_3 &= d_3 \\
\boldsymbol{n}_4 \cdot \boldsymbol{R} + s_4 &= d_4 \\
\boldsymbol{n}_5 \cdot \boldsymbol{R} + s_5 &= d_5 \\
&\vdots \\
\boldsymbol{n}_M \cdot \boldsymbol{R} + s_M &= d_M
\end{aligned}
$$

or

$$\begin{bmatrix} \cdots n_1 \cdots 1\,0\,0\,0\,0 \cdots 0 \\ \cdots n_2 \cdots 0\,1\,0\,0\,0 \cdots 0 \\ \cdots n_3 \cdots 0\,0\,1\,0\,0 \cdots 0 \\ \cdots n_4 \cdots 0\,0\,0\,1\,0 \cdots 0 \\ \cdots n_5 \cdots 0\,0\,0\,0\,1 \cdots 0 \\ \qquad \ddots \qquad \ddots \\ \cdots n_M \cdots 0\,0\,0\,0\,0 \cdots 1 \end{bmatrix} \begin{bmatrix} x \\ y \\ z \\ s_1 \\ s_2 \\ s_3 \\ s_4 \\ s_5 \\ \vdots \\ s_M \end{bmatrix} = \begin{bmatrix} d_1 \\ d_2 \\ d_3 \\ d_4 \\ d_5 \\ \vdots \\ d_M \end{bmatrix} \qquad (2.40)$$

with nonnegativity constraints imposed on the slack variables:

$$s_1 \geq 0, s_2 \geq 0, \ldots, s_M \geq 0.$$

Let's write this system in *augmented matrix* form. On occasion we'll include flags to keep track of which variable is associated with each column:

$$\begin{bmatrix} \cdots n_1 \cdots 1\,0\,0\,0\,0 \cdots 0 \mid d_1 \\ \cdots n_2 \cdots 0\,1\,0\,0\,0 \cdots 0 \mid d_2 \\ \cdots n_3 \cdots 0\,0\,1\,0\,0 \cdots 0 \mid d_3 \\ \cdots n_4 \cdots 0\,0\,0\,1\,0 \cdots 0 \mid d_4 \\ \cdots n_5 \cdots 0\,0\,0\,0\,1 \cdots 0 \mid d_5 \\ \qquad \ddots \qquad \ddots \\ \cdots n_M \cdots 0\,0\,0\,0\,0 \cdots 1 \mid d_M \end{bmatrix} \quad \text{(\textit{M} rows, \textit{M}+3+1 columns).} \qquad (2.41)$$

$$\underbrace{\qquad\qquad\qquad\qquad\qquad\qquad}_{x\ \ y\ \ z\ \ s_1\,s_2\,s_3\,s_4\,s_5\,\cdots\ s_M}$$

Now we have assumed that the diamond has a corner $\mathbf{R} = x\mathbf{i} + y\mathbf{j} + z\mathbf{k}$ where the first three constraints are active ($s_1 = s_2 = s_3 = 0$). Then the first 3-by-3 block in the matrix above would be the submatrix N addressed in Eq. (2.26) of Sect. 2.4:

$$\begin{bmatrix} \cdots n_1 \cdots 1\,0\,0\,0\,0 \cdots 0 \mid d_1 \\ \cdots n_2 \cdots 0\,1\,0\,0\,0 \cdots 0 \mid d_2 \\ \cdots n_3 \cdots 0\,0\,1\,0\,0 \cdots 0 \mid d_3 \\ \hline \cdots n_4 \cdots 0\,0\,0\,1\,0 \cdots 0 \mid d_4 \\ \cdots n_5 \cdots 0\,0\,0\,0\,1 \cdots 0 \mid d_5 \\ \qquad \ddots \qquad \ddots \\ \cdots n_M \cdots 0\,0\,0\,0\,0 \cdots 1 \mid d_M \end{bmatrix} \equiv \begin{bmatrix} \quad 1\,0\,0\,0\,0 \cdots 0 \mid d_1 \\ N \quad 0\,1\,0\,0\,0 \cdots 0 \mid d_2 \\ \quad 0\,0\,1\,0\,0 \cdots 0 \mid d_3 \\ \hline \cdots n_4 \cdots 0\,0\,0\,1\,0 \cdots 0 \mid d_4 \\ \cdots n_5 \cdots 0\,0\,0\,0\,1 \cdots 0 \mid d_5 \\ \qquad \ddots \qquad \ddots \\ \cdots n_M \cdots 0\,0\,0\,0\,0 \cdots 1 \mid d_M \end{bmatrix} \qquad (2.42)$$

$$\underbrace{\qquad\qquad\qquad\qquad}_{x\ \ y\ \ z\ \ s_1\,s_2\,s_3\,s_4\,s_5\,\cdots\ s_M} \qquad \underbrace{\qquad\qquad\qquad\qquad}_{x\ \ y\ \ z\ \ s_1\,s_2\,s_3\,s_4\,s_5\,\cdots\ s_M}$$

We can identify some identity and zero submatrices, of various sizes, in the augmented matrix[15]:

$$
\begin{array}{|ccc|ccc|}
 & |\ 3\ columns & |\ 3\ columns & |\ (M-3)\ columns & |
\end{array}
$$

$$
\begin{bmatrix}
 & | & & | & & | & d_1 \\
N & | & I & | & 0 & | & d_2 \\
 & | & & | & & | & d_3 \\
--- & | & --- & | & -\ ---\ - & | & \\
\cdots n_4 \cdots & | & & | & & | & d_4 \\
\cdots n_5 \cdots & | & & | & & | & d_5 \\
\vdots & | & 0 & | & I & | & \vdots \\
\vdots & | & & | & & | & \vdots \\
\cdots n_M \cdots & | & & | & & | & d_M
\end{bmatrix}.^{19}
\tag{2.43}
$$

Let us ask what would happen if we applied pivoting to (2.43) so as to *row-reduce* the submatrix N to a 3-by-3 identity matrix and (reduce the rows below it to zeros). (Typically we would pivot three times down the diagonal of N, although we may need to rearrange the first three rows to avoid nuisance zeros in the pivot positions.) We would end up with an augmented matrix whose system of equations is equivalent to the original constraints. What would it look like?

$$
\begin{bmatrix}
 & | & ? & ? & ? & |\ ? & \cdots & ? & | & ? \\
I & | & ? & ? & ? & |\ ? & \cdots & ? & | & ? \\
 & | & ? & ? & ? & |\ ? & \cdots & ? & | & ? \\
--- & | & - & - & - & - & - & | & - \\
 & | & ? & ? & ? & |\ ? & \cdots & ? & | & ? \\
 & | & ? & ? & ? & |\ ? & \cdots & ? & | & ? \\
0 & | & \vdots & \vdots & \vdots & |\ \vdots & \vdots & \vdots & | & \vdots \\
 & | & \vdots & \vdots & \vdots & |\ \vdots & \vdots & \vdots & | & ? \\
 & | & ? & ? & ? & |\ ? & \cdots & ? & | & ?
\end{bmatrix}
\tag{2.44}
$$

Recall (Sect. 2.3) that one feature of the augmented matrix formulation is based on the fact that when we append a column vector v to the right of a square (invertible) matrix N and row-reduce N to an identity, the vector v is reduced to $N^{-1}v$. Since the 3-by-3 submatrix in (2.43) to the right of N is the identity $I_{3\times 3}$, it's columns are reduced to $N^{-1}I = N^{-1}$, and (2.44) is actually

---

[15] It is worth noting that the whole $M \times M$ identity matrix resides in columns 4 to $M + 3$.

$$
\begin{bmatrix}
 & | & & & | & ? & ? & ? & | & ? \\
I & | & & N^{-1} & | & ? & ? & ? & | & ? \\
 & | & & & | & ? & ? & ? & | & ? \\
-\,-\,- & | & -\,-\,- & & | & -\,-\,-\,-\,-\,- & | & - \\
 & | & ? & ? & ? & | & ? & ? & ? & | & ? \\
 & | & ? & ? & ? & | & ? & ? & ? & | & ? \\
0 & | & & \vdots & & | & \ddots & & | & \vdots \\
 & | & ? & ? & ? & | & ? & ? & ? & | & ? \\
 & | & ? & ? & ? & | & ? & ? & ? & | & ?
\end{bmatrix}
\tag{2.45}
$$

Furthermore the rightmost 3-by-1 column is reduced to the solution of the first three equations represented by (2.42), namely the corner coordinates $R = [x\,y\,z]^T = N^{-1}[d_1\,d_2\,d_3]^T$ (remember we have assumed the first three slack variables are zero); and the intervening rows of zeros remain zeros:

$$
\begin{bmatrix}
 & | & & & | & & & | & x \\
I & | & & N^{-1} & | & & 0 & | & y \\
 & | & & & | & & & | & z \\
-\,-\,- & | & -\,-\,- & & | & -\,-\,-\,-\,-\,- & | & - \\
 & | & ? & ? & ? & | & ? & ? & ? & | & ? \\
 & | & ? & ? & ? & | & ? & ? & ? & | & ? \\
0 & | & & \vdots & & | & \ddots & & | & \vdots \\
 & | & ? & ? & ? & | & ? & ? & ? & | & ? \\
 & | & ? & ? & ? & | & ? & ? & ? & | & ?
\end{bmatrix}
\tag{2.46}
$$

Now in the first 3 columns, the lower rows were "zeroed out" by subtracting multiples of the rows of N; generically speaking, the row vector n_i ($i \geq 4$) was "annihilated" by a linear combination of the rows n_1, n_2, and n_3.

Question 11. How do we express a linear combination of the rows of a matrix?

Answer. If p, q, and r are the coefficients in the combination, then the linear combination given by the matrix product

$$
[p\,q\,r]\begin{bmatrix}
\cdots n_1 \cdots \\
\cdots n_2 \cdots \\
\cdots n_3 \cdots
\end{bmatrix} = [p\,q\,r][N]
\tag{2.47}
$$

does the trick. (Multiply out an example, if you don't see it.) (Similarly, the product

$$[N]\begin{bmatrix} p \\ q \\ r \end{bmatrix} = \begin{bmatrix} | & | & | \\ | & | & | \\ \downarrow & \downarrow & \downarrow \end{bmatrix}\begin{bmatrix} p \\ q \\ r \end{bmatrix} \tag{2.48}$$

gives the linear combination of the *columns.*)

Question 12. Which particular linear combination of the rows of N will annihilate the vector n_i?

Answer. We want

$$n_i - p_i n_1 - q_i n_2 - r_i n_3 = n_i - [p_i \ q_i \ r_i][N]$$

to become the row vector zero. Thus, solving for the coefficients, we get

$$n_i - [p_i \ q_i \ r_i][N] = 0 \quad or \quad [p_i \ q_i \ r_i] = n_i N^{-1}. \tag{2.49}$$

(No surprise here; $n_i - n_i N^{-1} N = 0$.) So the row-reduction step, which annihilates the last $M - 3$ rows (in columns 1–3), can be expressed in matrix language as

$$\begin{bmatrix} n_4 \\ n_5 \\ \vdots \\ n_M \end{bmatrix} - \begin{bmatrix} p_4 \ q_4 \ r_4 \\ p_5 \ q_5 \ r_5 \\ \vdots \\ p_M \ q_M \ r_M \end{bmatrix}[N] = 0 \text{ where } \begin{bmatrix} p_4 \ q_4 \ r_4 \\ p_5 \ q_5 \ r_5 \\ \vdots \\ p_M \ q_M \ r_M \end{bmatrix} = \begin{bmatrix} n_4 \\ n_5 \\ \vdots \\ n_M \end{bmatrix}[N]^{-1}. \tag{2.50}$$

The lower $M - 3$ rows in the *second* three columns of the augmented matrix (2.43), which *were* zeros, undergo subtraction of the *same* multiples ($[p_i \ q_i \ r_i]$) of the 3 rows above them—*which were the rows of the identity* $I_{3\times3}$. Thus these rows become

$$\begin{bmatrix} 0 \\ 0 \\ \vdots \\ 0 \end{bmatrix} - \begin{bmatrix} p_4 \ q_4 \ r_4 \\ p_5 \ q_5 \ r_5 \\ \vdots \\ p_M \ q_M \ r_M \end{bmatrix}[I] = \begin{bmatrix} 0 \\ 0 \\ \vdots \\ 0 \end{bmatrix} - \begin{bmatrix} n_4 \\ n_5 \\ \vdots \\ n_M \end{bmatrix}[N]^{-1}[I] = \begin{bmatrix} 0 \\ 0 \\ \vdots \\ 0 \end{bmatrix} - \begin{bmatrix} n_4 \\ n_5 \\ \vdots \\ n_M \end{bmatrix}[N]^{-1}$$

$$= - \begin{bmatrix} n_4 \\ n_5 \\ \vdots \\ n_M \end{bmatrix}\left[N_{col1}^{-1} \ N_{col2}^{-1} \ N_{col3}^{-1} \right]. \tag{2.51}$$

But these numbers are precisely the $t_i^{(K)}$'s defined in Eq. (2.34) of Sect. 2.4(!), and our matrix (2.46) above is further revealed to be

$$
\begin{array}{cccc}
\overbrace{\,|\,3\ columns\,|}^{} & \overbrace{\,3\ columns}^{} & |\quad (M-3)\ columns & | \\
\end{array}
$$

$$
\left[
\begin{array}{c|c|c|c}
I & N^{-1} & 0 & \begin{array}{c} x \\ y \\ z \end{array} \\
\hline
0 & -\begin{bmatrix} n_4 \\ n_5 \\ \vdots \\ n_M \end{bmatrix}[N]^{-1} & ? & ?
\end{array}
\right]
\tag{2.52}
$$

or

$$
\left[
\begin{array}{c|c|cccc}
\overbrace{|\,3\ columns}^{} & \overbrace{|\quad 3\ columns}^{} & |\quad (M-3)\ columns & & | \\
I & N^{-1} & 0 & & \begin{array}{c} x \\ y \\ z \end{array} \\
\hline
0 & \begin{array}{ccc} t_4^{(1)} & t_4^{(2)} & t_4^{(3)} \\ t_5^{(1)} & t_5^{(2)} & t_5^{(3)} \\ \vdots & & \\ \vdots & & \\ t_M^{(1)} & t_M^{(2)} & t_M^{(3)} \end{array} & \begin{array}{cccc} ? & ? & ? & ? \\ ? & ? & ? & ? \\ & \ddots & & \vdots \\ ? & ? & ? & ? \\ ? & ? & ? & ? \end{array}
\end{array}
\right].
$$

The lower rows of the next $(M-3)$ columns, which originally constituted an identity matrix in (2.43), are unchanged because the pivoting subtracts multiples of the zeros in the first three rows:

$$
\left[
\begin{array}{c|c|c|c}
I & N^{-1} & 0 & \begin{array}{c} x \\ y \\ z \end{array} \\
\hline
0 & -\begin{bmatrix} n_4 \\ n_5 \\ \vdots \\ n_M \end{bmatrix}[N]^{-1} & I & ?
\end{array}
\right]
\tag{2.53}
$$

Finally, when the row-reduction step (2.50), (2.51) is applied to the last $(M-3)$ d_j's in the right hand column of (2.43), the result is the slack variables (recall Eq. (2.40))[16]:

---

[16] As a check: if any of s_i entries are negative, then the first 3 constraints do *not* define a corner when they are active. We made a mistake in choosing them (and must resort to the strategy of Sect. 2.10).

$$
\begin{bmatrix} d_4 \\ d_5 \\ \vdots \\ d_M \end{bmatrix} - \begin{bmatrix} p_4 \; q_4 \; r_4 \\ p_5 \; q_5 \; r_5 \\ \vdots \\ p_M \; q_M \; r_M \end{bmatrix} \begin{bmatrix} d_1 \\ d_2 \\ d_3 \end{bmatrix} = \begin{bmatrix} d_4 \\ d_5 \\ \vdots \\ d_M \end{bmatrix} - \begin{bmatrix} n_4 \\ n_5 \\ \vdots \\ n_M \end{bmatrix} [N]^{-1} \begin{bmatrix} d_1 \\ d_2 \\ d_3 \end{bmatrix}
$$

$$
= \begin{bmatrix} d_4 \\ d_5 \\ \vdots \\ d_M \end{bmatrix} - \begin{bmatrix} n_4 \\ n_5 \\ \vdots \\ n_M \end{bmatrix} \begin{bmatrix} x \\ y \\ z \end{bmatrix} = \begin{bmatrix} d_4 \\ d_5 \\ \vdots \\ d_M \end{bmatrix} - \begin{bmatrix} n_4 \\ n_5 \\ \vdots \\ n_M \end{bmatrix} [R] = \begin{bmatrix} s_4 \\ s_5 \\ \vdots \\ s_M \end{bmatrix}.
$$

$$
\tag{2.54}
$$

Wow! *Row-reduction of the first three columns of the augmented matrix* (2.41) *gives us an equivalent expression of the constraint equations with the format*

$$
\begin{bmatrix}
 & | & & | & & | & x \\
I & | & N^{-1} & | & 0 & | & y \\
 & | & & | & & | & z \\
- - - & | & - \; - \; - & | & - - - - - & | & - \\
 & | & t_4^{(1)} t_4^{(2)} t_4^{(3)} & | & & | & s_4 \\
 & | & t_5^{(1)} t_5^{(2)} t_5^{(3)} & | & & | & s_5 \\
0 & | & \vdots & | & I & | & \vdots \\
 & | & \vdots & | & & | & \vdots \\
 & | & t_M^{(1)} t_M^{(2)} t_M^{(3)} & | & & | & s_M
\end{bmatrix}
\tag{2.55}
$$

or[17]

$$
\begin{bmatrix}
 & | & & | & & | & x \\
I & | & N^{-1} & | & 0 & | & y \\
 & | & & | & & | & z \\
- - - & | & - \; - \; - & | & - - - - - & | & \\
 & | & \begin{bmatrix} n_4 \\ n_5 \\ \vdots \\ n_M \end{bmatrix} & | & & | & \begin{matrix} s_4 \\ s_5 \\ \vdots \\ s_M \end{matrix} \\
0 & | & - \qquad\quad [N]^{-1} & | & I & | & \\
\end{bmatrix}.
$$

It generates almost all of the data—the s's, the t's, the edge/columns in N^{-1} — that we need to proceed with the corner update. All that's missing is the calculation $-c[N]^{-1}$ in Eq. (2.33) of Sect. 2.4, which determines the particular edge along which we progress.

---

[17] Note that the first three columns of the $M \times M$ identity reside in columns 1–3, while the remaining $M - 3$ columns reside in columns 7 to $M + 3$.

We'll show that $-cN^{-1}$ can easily be obtained by appending a final row to the augmented matrix (2.41), containing the coefficient vector c of the objective function $c \cdot R$ and padded with zeros:

$$
\begin{bmatrix}
\cdots n_1 \cdots 1\ 0\ 0\ 0\ 0\ \cdots 0 & d_1 \\
\cdots n_2 \cdots 0\ 1\ 0\ 0\ 0\ \cdots 0 & d_2 \\
\cdots n_3 \cdots 0\ 0\ 1\ 0\ 0\ \cdots 0 & d_3 \\
\cdots n_4 \cdots 0\ 0\ 0\ 1\ 0\ \cdots 0 & d_4 \\
\cdots n_5 \cdots 0\ 0\ 0\ 0\ 1\ \cdots 0 & d_5 \\
\qquad \ddots \qquad\qquad \ddots & \\
\cdots n_M \cdots 0\ 0\ 0\ 0\ 0\ \cdots 1 & d_M \\
\cdots c \cdots\ \ 0\ 0\ 0\ 0\ 0\ \cdots 0 & 0
\end{bmatrix} . \qquad (2.56)
$$

This matrix and its row-reduced versions are known as *Simplex tableaux*.[18]

When c (at the bottom of the first three columns) is zeroed out by subtracting the appropriate multiples $[p\ q\ r]$ of the rows of N in the manner of Eq. (2.49),

$$[c] - [p\ q\ r][N] = [0],$$

the multiples are given by

$$[p\ q\ r] = [c][N]^{-1};$$

thus data in this "c-row" undergo the modifications depicted below:

$$subtract\ [p\ q\ r] \equiv c[N]^{-1} times \left[N\ \ I_{3\times 3}\ \ \mathbf{0}_{3\times(M-3)}\ \begin{bmatrix} d_1 \\ d_2 \\ d_3 \end{bmatrix} \right]$$

$$from\ [\cdots c \cdots 00000 \cdots 00], \quad yielding$$

$$\left[0\ 0\ 0\ \ -c[N]^{-1}[I]\ \ -c[N]^{-1}[\mathbf{0}]\ \ -cN^{-1} \begin{bmatrix} d_1 \\ d_2 \\ d_3 \end{bmatrix} \right]$$

or

$$\left[0\ 0\ 0\ \ -c[N]^{-1}\ \ \mathbf{0}\ \ -c \cdot R \right].$$

---

[18] (plural of *tableau*).

And we have $-cN^{-1}$, the remaining data we need in order to select the edge(s) which increase the objective! As an added bonus, the final entry of the matrix, $-c \cdot R$, gives us the (negative of the) value of the objective function at the diamond corner.

A little nomenclature will be helpful. The tableau (2.56), which contains the constraint equations and the objective function coefficients, will be designated as the **constraint tableau**:

$$\begin{bmatrix}
\cdots n_1 \cdots & 1\,0\,0\,0\,0 & \cdots\ 0 & d_1 \\
\cdots n_2 \cdots & 0\,1\,0\,0\,0 & \cdots\ 0 & d_2 \\
\cdots n_3 \cdots & 0\,0\,1\,0\,0 & \cdots\ 0 & d_3 \\
\cdots n_4 \cdots & 0\,0\,0\,1\,0 & \cdots\ 0 & d_4 \\
\cdots n_5 \cdots & 0\,0\,0\,0\,1 & \cdots\ 0 & d_5 \\
 & \ddots & \ddots & \\
\cdots n_M \cdots & 0\,0\,0\,0\,0 & \cdots\ 1 & d_M \\
\cdots c \cdots & 0\,0\,0\,0\,0 & \cdots\ 0 & 0
\end{bmatrix} . \text{(2.56 repeated)}$$

The tableau resulting from the pivoting will be designated as the **row-reduced tableau**:

$$\begin{bmatrix}
 & \Big| & & \Big| & & \Big| & x \\
I & \Big| & N^{-1} & \Big| & 0 & \Big| & y \\
 & \Big| & & \Big| & & \Big| & z \\
- - - & \Big| & - - - & \Big| & - - - - & \Big| & - \\
 & \Big| & t_4^{(1)}\ t_4^{(2)}\ t_4^{(3)} & \Big| & & \Big| & s_4 \\
 & \Big| & t_5^{(1)}\ t_5^{(2)}\ t_5^{(3)} & \Big| & & \Big| & s_5 \\
0 & \Big| & \vdots & \Big| & I & \Big| & \vdots \\
 & \Big| & \vdots & \Big| & & \Big| & \vdots \\
 & \Big| & t_M^{(1)}\ t_M^{(2)}\ t_M^{(3)} & \Big| & & \Big| & s_M \\
- - - & \Big| & - - - & \Big| & - - - - & \Big| & - \\
0\ 0\ 0 & \Big| & -cN^{-1} & \Big| & 0\ \cdots\ 0 & \Big| & -c \cdot R
\end{bmatrix} . \quad (2.57)$$

In both tableaux, the bottom row will be designated as the c-row, and the submatrix lying above it (which is equivalent to the augmented matrix for the constraint equations) will be the *constraint submatrix*.

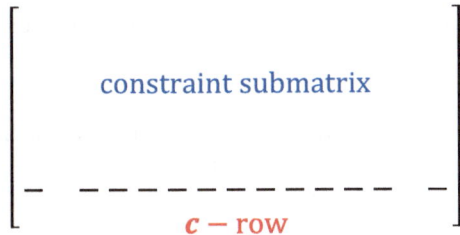

Note the following:

Characteristics of the Row-Reduced Tableau

(a) Its constraint submatrix contains all the columns of the M by M identity matrix.
(b) Its c-row contains zeros in all columns except the last, and those corresponding to the active slack variables.
(c) When $[x\ y\ z]$ locates a legitimate corner, the entries in the last column corresponding to the inactive slack variables are nonnegative.

Let's summarize.

Row Reduction Step for the Simplex Method

1. List the constraint inequalities with the constraints that are active at a particular corner of the diamond listed first.
2. Reformulate the inequalities as equalities by inserting nonnegative slack variables; express these equations as an augmented matrix.
3. Append a row containing the objective vector c padded with zeros, completing the constraint tableau (2.56).
4. Perform pivoting (with row exchanges if necessary) on this tableau so that its upper left 3-by-3 submatrix is row-reduced to the identity.

Then the parameters $-cN^{-1}$, $t_i^{(j)}$, and s_i necessary for selecting the constraints that specify the next corner[19] can be read from the resulting row-reduced tableau in compliance with (2.57). Rearrange the constraint list accordingly and start over.

Note: the user-friendly organization of the parameter display in the row-reduced tableau (2.57) is premised on the specific format of the constraint tableau (2.56). "Starting over" from a different corner will entail rearranging constraint rows, rearranging columns, and renumbering the constraints and slack variables to rigidly conform to format (2.56); this will be demonstrated in our examples. (Fortunately such fussiness is obviated in the sleek version of the algorithm presented in Sects. 2.8–2.9.)

---

[19] (or terminate the computation).

Before we explore ways to streamline the Simplex algorithm, let's work a simple problem with this formulation.

Example 2.1 Maximize the objective function z in the polyhedron defined by the four constraints $x \geq 0$, $y \geq 0$, $z \geq 0$, $x + y + z \leq 1$.

Solution. The polyhedron is displayed in Fig. 2.13, and the solution is obvious: the maximum z is one, achieved at the corner $(0, 0, 1)$. But let's see how the algorithm derives this if we start from the corner at $(1, 0, 0)$; it should direct us along the indicated edge, then tell us we are at the maximum.

Expressing the constraints as equations with nonnegative slack variables and identifying those that are active at the corner $(1, 0, 0)$ we have

$$\begin{aligned}
-x + s_1 &= 0 & inactive\ (s_1 = 1) \\
-y + s_2 &= 0 & active\ (s_2 = 0) \\
-z + s_3 &= 0 & active\ (s_3 = 0) \\
x + y + z + s_4 &= 1 & active\ (s_4 = 0)
\end{aligned} \tag{2.58}$$

But since we are listing the active constraints first, we reorder Eq. (2.58) and, accordingly, renumber the slack variables to write them as

Fig. 2.13 Example 2.1 polyhedron

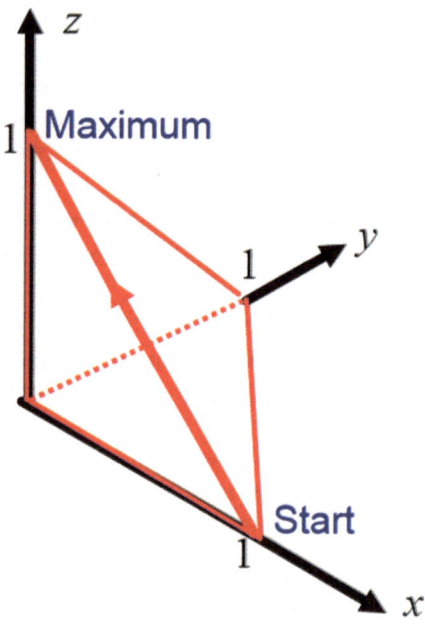

$$
\begin{aligned}
&(constraint\ \#1) & -y + s_1 &= 0 & active\ (s_1 = 0)\\
&(constraint\ \#2) & -z + s_2 &= 0 & active\ (s_2 = 0)\\
&(constraint\ \#3)\ x + y + z + s_3 &= 1 & active\ (s_3 = 0)\\
&(constraint\ \#4) & -x + s_4 &= 0 & inactive\ (s_4 = 1).
\end{aligned}
\tag{2.59}
$$

Since the objective function $c \cdot R$ equals z, c equals $[0\,0\,1]$ or k. Forming the augmented matrix, appending a row containing the zero-padded c, *flagging the slack variables and the constraints to preserve our sanity,* and partitioning the matrix, we display the constraint tableau as

$$
\begin{bmatrix}
0 & -1 & 0 & | & 1 & 0 & 0 & | & 0 & | & 0 \\
0 & 0 & -1 & | & 0 & 1 & 0 & | & 0 & | & 0 \\
1 & 1 & 1 & | & 0 & 0 & 1 & | & 0 & | & 1 \\
- & - & - & - & - & - & - & - & - & - & - \\
-1 & 0 & 0 & | & 0 & 0 & 0 & | & 1 & | & 0 \\
- & - & - & - & - & - & - & - & - & - & - \\
0 & 0 & 1 & | & 0 & 0 & 0 & | & 0 & | & 0
\end{bmatrix}
\begin{matrix}
\#1 \\ \#2 \\ \#3 \\ \\ \#4 \\ \\ (c)
\end{matrix}
\tag{2.60}
$$

$$
\underbrace{}\ \ x\quad y\quad z\quad\ \ s_1\ s_2\ s_3\quad s_4
$$

(consistent with (2.56)). (Remember the first row of (2.60) is a shorthand for

$$
0x - 1y + 0z + 1s_1 + 0s_2 + 0s_3 + 0s_s = 0,\ \text{etc.)}
$$

To find the $-cN^{-1}$, t, and s parameters that tell us how to get to the next corner, we row-reduce the first three columns: that is, we pivot 3 times to install an identity in the first 3-by-3 submatrix and zero out the rows below it. (Recall from Sect. 2.3 that a pivoted constraint equation still enforces the same constraint—"Pivoted Content Immunity".) The result is the row-reduced tableau

$$
\begin{bmatrix}
1 & 0 & 0 & | & 1 & 1 & 1 & | & 0 & | & 1 \\
0 & 1 & 0 & | & -1 & 0 & 0 & | & 0 & | & 0 \\
0 & 0 & 1 & | & 0 & -1 & 0 & | & 0 & | & 0 \\
- & - & - & - & - & - & - & - & - & - & - \\
0 & 0 & 0 & | & 1 & 1 & 1 & | & 1 & | & 1 \\
- & - & - & - & - & - & - & - & - & - & - \\
0 & 0 & 0 & | & 0 & 1 & 0 & | & 0 & | & 0
\end{bmatrix}
\begin{matrix}
\#1 \\ \#2 \\ \#3 \\ \\ \#4 \\ \\ (c)
\end{matrix}
\tag{2.61}
$$

$$
x\quad y\quad z\quad\ \ s_1\ s_2\ s_3\quad s_4
$$

Note that for the parameters of this problem the row-reduced tableau template (2.57) is

$$
(25) \quad
\begin{bmatrix}
& \vdots & & \vdots & & \vdots & x \\
I & \vdots & N^{-1} & \vdots & 0 & \vdots & y \\
& \vdots & & \vdots & & \vdots & z \\
\text{---} & \text{---} & \text{---} & \text{---} & \text{---} & \text{---} & \text{---} \\
0 & \vdots & t_4^{(1)}\, t_4^{(2)}\, t_4^{(3)} & \vdots & 1 & \vdots & s_4 \\
\text{---} & \text{---} & \text{---} & \text{---} & \text{---} & \text{---} & \text{---} \\
0\ 0\ 0 & \vdots & -cN^{-1} & \vdots & 0 & \vdots & -c\cdot R
\end{bmatrix}
\cdot \quad (2.62)
$$

$$x\ y\ z \qquad s_1\ \ s_2\ \ s_3 \qquad\qquad s_4$$

Let's compare (2.61) with (2.62).

(i)   The first upper 3-by-3 submatrix is the identity, and the rows below it contain zeros. 👍

(ii) The second upper 3-by-3 submatrix is the inverse of the first submatrix in (2.60). (Mental arithmetic confirms this.) 👍

(iii) The seventh column is comprised of a 3-by-1 zero submatrix, a 1-by-1 identity, and a zero in the last row. 👍

(iv) The first three entries of the final column are the current corner coordinates $x = 1$, $y = 0$, $z = 0$. It's final entry is the value of the objective function $-c \cdot R = -z$ at this corner: $z = 0$. 👍

(v)   In the c-row the second 1-by-3 submatrix $[0\ 1\ 0]$ equals $-cN^{-1}$; the positive 1 in the second position of this submatrix (above the symbol s_2, i. e. the fifth column of the tableau) indicates that we should move along the edge of the diamond corresponding to the negative of the second column ($K = 2$) of the inverse submatrix N^{-1} at the top (recall Eq. (2.33, Sect. 2.4)): namely, in the direction $-[1\ \ 0\ -1]^T$ or $-i + k$ (see Fig. 2.13). This will deactivate the second constraint in (2.58)—namely, $-z + s_2 = 0$; so s_2 will increase. The second row is fated to be shifted out of the group of active constraints (the first three rows). *To summarize: the location of the* 1 *in the bottom row tells us to conduct the s/t competition in the* s_2 *column.* 👍

(vi) Since $M = 4$ constraints, the $t_j^{(K=2)}$'s all reside in the *fourth* row: $j = 4$. So there is only one contestant $t_j^{(2)}$ in the column of interest ($t_4^{(2)} = 1$, in column 5 of the tableau); it is positive, and the ratio $s_4/t_4^{(2)}$ is $1/1$. Thus Eq. (2.36, Sect. 2.4) says w equals 1, and Eq. (2.37) of that section says the next corner to be selected will be at 👍

$$
R + \delta R = R - wN^{-1}_{colK} = R - w
\begin{bmatrix}
n_1 \to \\
n_2 \to \\
n_3 \to
\end{bmatrix}^{-1}
\begin{bmatrix}
\vert \\
e_2 \\
\downarrow
\end{bmatrix}
$$

$$
= \begin{bmatrix} 1 \\ 0 \\ 0 \end{bmatrix} - (1) \begin{bmatrix} 1 & 1 & 1 \\ -1 & 0 & 0 \\ 0 & -1 & 0 \end{bmatrix} \begin{bmatrix} 0 \\ 1 \\ 0 \end{bmatrix} = \begin{bmatrix} 1 \\ 0 \\ 0 \end{bmatrix} - (1) \begin{bmatrix} 1 \\ 0 \\ -1 \end{bmatrix} = \begin{bmatrix} 0 \\ 0 \\ 1 \end{bmatrix}
$$

$$\tag{2.63}$$

(consistent with Fig. 2.13). It will activate the *fourth* constraint in (2.60) (s_4 will become 0 in the equation $-x + s_4 = 0$). *In brief, the s/t winner 1 lies in the fourth row; the fourth constraint will be activated.*

Synopsis: The positive **1** and the s/t winner **1** in the tableau

$$
\begin{bmatrix}
1 & 0 & 0 & | & 1 & 1 & 1 & | & 0 & | & 1 \\
0 & 1 & 0 & | & -1 & 0 & 0 & | & 0 & | & 0 \\
0 & 0 & 1 & | & 0 & -1 & 0 & | & 0 & | & 0 \\
- & - & - & - & - & - & - & - & - & - & - \\
0 & 0 & 0 & | & 1 & \mathbf{1} & 1 & | & 1 & | & 1 \\
- & - & - & - & - & - & - & - & - & - & - \\
0 & 0 & 0 & | & 0 & \mathbf{1} & 0 & | & 0 & | & 0
\end{bmatrix}
\begin{matrix}
\#1 \\ \#2 \\ \#3 \\ \\ \#4 \\ \\ (c)
\end{matrix}
\quad . \text{(2.61 repeated)}
$$

$$
\begin{matrix}
x & y & z & s_1 & s_2 & s_3 & s_4
\end{matrix}
$$

direct us to exchange rows **2** and **4** in the constraint tableau.

With this exchange, tableau (2.60) becomes

$$
\begin{bmatrix}
0 & -1 & 0 & | & 1 & 0 & 0 & | & 0 & | & 0 \\
-1 & 0 & 0 & | & 0 & 0 & 0 & | & 1 & | & 0 \\
1 & 1 & 1 & | & 0 & 0 & 1 & | & 0 & | & 1 \\
- & - & - & - & - & - & - & - & - & - & - \\
0 & 0 & -1 & | & 0 & 1 & 0 & | & 0 & | & 0 \\
- & - & - & - & - & - & - & - & - & - & - \\
0 & 0 & 1 & | & 0 & 0 & 0 & | & 0 & | & 0
\end{bmatrix}
\begin{matrix}
\#1 \\ \#4 \\ \#3 \\ \\ \#2 \\ \\ (c)
\end{matrix}
\quad .
$$

$$\tag{2.64}$$

$$
\begin{matrix}
x & y & z & s_1 & s_2 & s_3 & s_4
\end{matrix}
$$

Before we carry out the pivoting to find the next corner, observe that (2.64) does not have the correct format (2.56) for initiating the tableau reduction. First, the identity columns are out of place. If we expect to find the $-cN^{-1}$, t, and s in the convenient positions indicated in the template (2.57), we need to switch the positions of the columns for s_2 and s_4.

Question 13. What is the significance of switching columns in an augmented matrix?

Answer. A simple example will make it clear. The system

$$\begin{cases} 1x + 2y + 3z = 4 \\ 5x + 6y + 7z = 8 \end{cases} \tag{2.65}$$

is represented by the augmented matrix

$$\begin{bmatrix} 1\ 2\ 3 \mid 4 \\ 5\ 6\ 7 \mid 8 \end{bmatrix}. \tag{2.66}$$

If we switch its second and third columns

$$\begin{bmatrix} 1\ 3\ 2 \mid 4 \\ 5\ 7\ 6 \mid 8 \end{bmatrix}, \tag{2.67}$$

it represents

$$\begin{cases} 1x + 3y + 2z = 4 \\ 5x + 7y + 6z = 8 \end{cases}, \tag{2.68}$$

an entirely different system with different solutions(!) Here we avoid this calamity by appending column labels to (2.66)

$$\underbrace{\begin{bmatrix} 1 & 2 & 3 & \mid & 4 \\ 5 & 6 & 7 & \mid & 8 \end{bmatrix}}_{x \quad y \quad z}, \tag{2.69}$$

and carrying each column's label with it in the exchange:

$$\underbrace{\begin{bmatrix} 1 & 3 & 2 & \mid & 4 \\ 5 & 7 & 6 & \mid & 8 \end{bmatrix}}_{x \quad z \quad y}. \tag{2.70}$$

Accordingly, we swap the columns for s_2 and s_4 in (2.64):

$$\underbrace{\begin{bmatrix} 0 & -1 & 0 & \mid & 1 & 0 & 0 & \mid & 0 & \mid & 0 \\ -1 & 0 & 0 & \mid & 0 & 1 & 0 & \mid & 0 & \mid & 0 \\ 1 & 1 & 1 & \mid & 0 & 0 & 1 & \mid & 0 & \mid & 1 \\ - & - & - & - & - & - & - & - & - & - & - \\ 0 & 0 & -1 & \mid & 0 & 0 & 0 & \mid & 1 & \mid & 0 \\ - & - & - & - & - & - & - & - & - & - & - \\ 0 & 0 & 1 & \mid & 0 & 0 & 0 & \mid & 0 & \mid & 0 \end{bmatrix}}_{x \quad y \quad z \quad\ s_1 \ \ s_4 \ \ s_3 \qquad s_2} \begin{matrix} \#1 \\ \#4 \\ \#3 \\ \\ \#2 \\ \\ (c) \end{matrix}. \tag{2.71}$$

You may recall that Sect. 2.3 states that rearranging columns has no effect on pivoting ("Column Location Immunity"); if we wish we can switch the column positions back after a pivot, and the results will be the same as if we had left the columns in place.

Now we renumber the constraints in (2.71) to attain conformance with the template (2.56):

$$
\left[\begin{array}{ccc|ccc|c|c}
0 & -1 & 0 & 1 & 0 & 0 & 0 & 0 \\
-1 & 0 & 0 & 0 & 1 & 0 & 0 & 0 \\
1 & 1 & 1 & 0 & 0 & 1 & 0 & 1 \\
\hline
0 & 0 & -1 & 0 & 0 & 0 & 1 & 0 \\
\hline
0 & 0 & 1 & 0 & 0 & 0 & 0 & 0
\end{array}\right]
\begin{array}{l}
\left.\begin{array}{l}\#1'\\\#2'\\\#3'\end{array}\right\}\\[2pt]
\#4'\\[2pt]
(c)
\end{array}
\qquad (2.72)
$$

$$\underbrace{}$$
$$\quad x \quad\; y \quad\; z \qquad s'_1 \;\; s'_2 \;\; s'_3 \qquad\quad s'_4$$

where $\left[\#1' \; \#2' \; \#3' \; \#4'\right] \equiv \left[\#1 \; \#4 \; \#3 \; \#2\right]$,
$\left[s'_1 \; s'_2 \; s'_3 \; s'_4\right] \equiv \left[s_1 \; s_4 \; s_3 \; s_2\right]$.[20]

Row-reducing the first three columns in (2.72), we uncover the $-cN^{-1}$, t, and s parameters for this corner,

$$
\left[\begin{array}{ccc|ccc|c|c}
1 & 0 & 0 & 0 & -1 & 0 & 0 & 0 \\
0 & 1 & 0 & -1 & 0 & 0 & 0 & 0 \\
0 & 0 & 1 & 1 & 1 & 1 & 0 & 1 \\
\hline
0 & 0 & 0 & 1 & 1 & 1 & 1 & 1 \\
\hline
0 & 0 & 0 & -1 & -1 & -1 & 0 & -1
\end{array}\right]
\begin{array}{l}
\left.\begin{array}{l}\#1'\\\#2'\\\#3'\end{array}\right\}\\[2pt]
\#4'\\[2pt]
(c)
\end{array}
\qquad, \qquad (2.73)
$$

$$\underbrace{}$$
$$\quad x \quad\; y \quad\; z \qquad s'_1 \;\; s'_2 \;\; s'_3 \qquad\quad s'_4$$

by comparing with (2.62):

$$
\left[\begin{array}{ccc|ccc|c|c}
 & I & & & N^{-1} & & 0 & x \\
\hline
 & 0 & & t'^{(1)}_4 & t'^{(2)}_4 & t'^{(3)}_4 & 1 & s'_4 \\
\hline
0 & 0 & 0 & & -cN^{-1} & & 0 & -c\cdot R
\end{array}\right]
\begin{array}{l}
\left.\begin{array}{l}\#1'\\\#2'\\\#3'\end{array}\right\}\\[2pt]
\#4'\\[2pt]
(c)
\end{array}
\quad.
$$

$$\underbrace{}$$
$$\quad x \;\; y \;\; z \qquad s'_1 \;\; s'_2 \;\; s'_3 \qquad\quad s'_4$$

The absence of positive entries in the $-cN^{-1}$ block indicates that this corner maximizes the objective; we are done. The coordinates $x = 0$, $y = 0$, $z = 1$ of this corner

---

[20] If you like, you could think about why the d's and the s's undergo the same reordering.

are given in the first 3 rows of the last column (and agree with (2.63)). The value of the objective (z) is the negative of the final entry of the matrix, $-(-1) = 1$. The only nonzero slack variable is $s_4' \equiv s_2 = 1$, consistent with $-z + s_2 = 0$ (Eq. (2.59)).

In summary: the maximum value of z in the diamond equals 1, occurring at $x = 0$, $y = 0$, $z = 1$. All in agreement with Fig. 2.13.

Three salient points to take from this section:

(i)   The search for the corner maximizing the objective function must initiate from a legitimate corner characterized by three active constraints. The legitimacy of the corner is reflected in the nonnegativity of the other s_i entries in the row-reduced matrix (2.57). (If row reduction reveals one of them to be negative, the point where the corresponding supporting planes intersect does not lie on the diamond.) Sect. 2.10 will explain how to find a legitimate corner when none is evident.

(ii) The data in the c-row of the row-reduced matrix provides the information needed to *increase the objective function*. They reveal which active constraints could be deactivated.

(iii) The s/t ratio competition ensures the continuance of the *nonnegativity of the s_i entries* in the next row-reduced matrix; that is, it dictates which inactive constraint should be activated in order that the next corner will also be legitimate.

2.7   A Two-Dimensional Example

Practitioners of the Simplex algorithm have tweaked it to make it more efficient, but before we turn to this matter we explore a second model. Example 2.2, depicted in Fig. 2.14, differs from Example 2.1 of Sect. 2.6 in that the diamond is two dimensional (to reduce the number of computations), but it has 5 constraints to fully exercise the scope of the algorithm.

We attack this problem using the clunky but reliable machinery of Sect. 2.6, which can be expressed in a nutshell as follows.

(i)   Write the constraints and objective in the format of the constraint tableau template (2.56), Sect. 2.6, tailored to our dimensions, with the active constraints—defining a known legitimate corner—listed first. The template is

Fig. 2.14 Maximize $x + 2y$ in the shaded pentagon

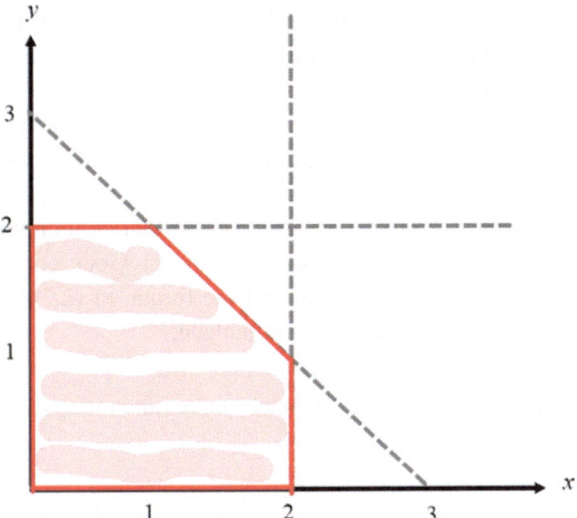

$$
\left[
\begin{array}{cc|cc|ccc|c}
\boldsymbol{n_1} \rightarrow & & & & & & & d_1 \\
\boldsymbol{n_2} \rightarrow & & \multicolumn{2}{c|}{\boldsymbol{I}_{2\times2}} & \multicolumn{3}{c|}{\boldsymbol{0}_{2\times3}} & d_2 \\
\hline
\boldsymbol{n_3} \rightarrow & & & & & & & d_3 \\
\boldsymbol{n_4} \rightarrow & & \multicolumn{2}{c|}{\boldsymbol{0}_{3\times2}} & \multicolumn{3}{c|}{\boldsymbol{I}_{3\times3}} & d_4 \\
\boldsymbol{n_5} \rightarrow & & & & & & & d_5 \\
\hline
\boldsymbol{c} \rightarrow & & 0 \quad 0 & & 0 \quad 0 \quad 0 & & & 0
\end{array}
\right]
\begin{array}{l} \text{\#1} \\ \text{\#2} \\ \\ \text{\#3} \\ \text{\#4} \\ \text{\#5} \\ \\ \end{array}
\qquad (2.74)
$$

$$
\begin{array}{ccccccc}
x & y & & s_1 & s_2 & s_3 \; s_4 \; s_5 &
\end{array}
$$

(ii) Apply pivoting to render the constraint tableau into the form of the row-reduced tableau (2.57), Sect. 2.6:

$$
\left[
\begin{array}{cc|cc|ccc|c}
\multicolumn{2}{c|}{\boldsymbol{I}_{2\times2}} & \multicolumn{2}{c|}{\boldsymbol{N}^{-1}} & \multicolumn{3}{c|}{\boldsymbol{0}_{2\times3}} & \begin{array}{c} x \\ y \end{array} \\
\hline
0 \quad 0 & & t_3^{(2)} & & & & & s_3 \\
0 \quad 0 & & t_4^{(1)} \; t_4^{(2)} & & \multicolumn{3}{c|}{\boldsymbol{I}_{3\times3}} & s_4 \\
0 \quad 0 & & t_5^{(1)} \; t_5^{(2)} & & & & & s_5 \\
\hline
0 \quad 0 & & \multicolumn{2}{c|}{-\boldsymbol{c}\boldsymbol{N}^{-1}} & 0 \quad 0 \quad 0 & & & -\boldsymbol{c} \cdot \boldsymbol{R}
\end{array}
\right]
\begin{array}{l} \text{\#1} \\ \text{\#2} \\ \\ \vdots \\ \\ \text{\#5} \\ \end{array}
\qquad (2.75)
$$

$$
\begin{array}{ccccccc}
x & y & & s_1 & s_2 & s_3 \; s_4 \; s_5 &
\end{array}
$$

(iii) If no entries in the $-cN^{-1}$ submatrix are positive, the maximum value of the objective function $c \cdot R$ occurs at this diamond corner, with coordinates x, y. The algorithm is terminated.

(iv) Otherwise, if no $t_i^{(j)}$ in the column of interest is positive, the polygon and the objective function are unbounded. The algorithm is terminated.

(v) Otherwise, pick a column $K (= 1 \text{ or } 2)$ with a positive entry in the $-cN^{-1}$ submatrix; pick a positive $t_{I(=3,4,5)}^{(K)}$ entry in that column with the lowest $s_I / t_I^{(K)}$ ratio.

(vi) Exchange rows I and K in the previous constraint tableau (2.74). Rearrange and renumber the data so that the format of (2.74) is recovered. Row reduce and repeat with this new constraint tableau.

Example 2.2 Maximize the objective function $x + 2y$ in the convex polygon defined by the constraints

$$x \geq 0, y \geq 0, x \leq 2, y \leq 2, x + y \leq 3 \qquad \text{(Fig. 2.14)}$$

Solution. We'll begin the search at the corner $x = y = 0$. Introducing slack variables we formulate the constraints as

$$\begin{cases} \text{\#1} & -x + s_1 = 0 & active \ (s_1 = 0) \\ \text{\#2} & -y + s_2 = 0 & active \ (s_2 = 0) \\ \text{\#3} & x + s_3 = 2 & inactive \ (s_3 = 0) \\ \text{\#4} & y + s_4 = 2 & inactive \ (s_4 = 0) \\ \text{\#5} \ x + y + s_5 = 3 & inactive \ (s_5 = 0). \end{cases} \qquad (2.76)$$

The objective function $c \cdot R = x + 2y$ so c equals $[1 \ 2]$. The constraint tableau (2.74) becomes

$$\left. \begin{bmatrix} -1 & 0 & | & 1 & 0 & | & 0 & 0 & 0 & | & 0 \\ 0 & -1 & | & 0 & 1 & | & 0 & 0 & 0 & | & 0 \\ - & - & - & - & - & - & - & - & - & - & - \\ 1 & 0 & | & 0 & 0 & | & 1 & 0 & 0 & | & 2 \\ 0 & 1 & | & 0 & 0 & | & 0 & 1 & 0 & | & 2 \\ 1 & 1 & | & 0 & 0 & | & 0 & 0 & 1 & | & 3 \\ - & - & - & - & - & | & - & - & - & - \\ 1 & 2 & | & 0 & 0 & | & 0 & 0 & 0 & | & 0 \end{bmatrix} \right\} \begin{matrix} \#1 \\ \#2 \\ - \\ \#3 \\ \#4 \\ \#5 \\ - \\ \\ \end{matrix} \qquad (2.77)$$

$$\quad\ x \quad\ y \qquad s_1 \quad s_2 \qquad s_3 \quad s_4 \quad s_5$$

By pivoting twice we row-reduce the upper left 2-by-2 submatrix to the identity:

$$
\left[
\begin{array}{cc|cc|ccc|c}
1 & 0 & -1 & 0 & 0 & 0 & 0 & 0 \\
0 & 1 & 0 & -1 & 0 & 0 & 0 & 0 \\
\hline
0 & 0 & 1 & 0 & 1 & 0 & 0 & 2 \\
0 & 0 & 0 & 1 & 0 & 1 & 0 & 2 \\
0 & 0 & 1 & 1 & 0 & 0 & 1 & 3 \\
\hline
0 & 0 & 1 & 2 & 0 & 0 & 0 & 0
\end{array}
\right]
\begin{array}{l}
\#1 \\ \#2 \\ \\ \#3 \\ \#4 \\ \#5 \\ \\ \
\end{array}
\qquad (2.78)
$$

$$
\begin{array}{cccccccc}
x & y & s_1 & s_2 & s_3 & s_4 & s_5 &
\end{array}
$$

Comparing (2.78) with the template (2.75), we see that the positive entries in the c-row indicate that we can increase the objective function by deactivating *either* of the first two constraints. We arbitrarily choose the first constraint, i.e. we'll deactivate s_1 by replacing the first row in the constraint tableau (2.77). The s/t competition winner for the s_1 column in tableau (2.78) occurs in row 3, so we'll exchange row 3 with row 1 in the constraint tableau (2.77):

$$
\left[
\begin{array}{cc|cc|ccc|c}
1 & 0 & 0 & 0 & 1 & 0 & 0 & 2 \\
0 & -1 & 0 & 1 & 0 & 0 & 0 & 0 \\
\hline
-1 & 0 & 1 & 0 & 0 & 0 & 0 & 0 \\
0 & 1 & 0 & 0 & 0 & 1 & 0 & 2 \\
1 & 1 & 0 & 0 & 0 & 0 & 1 & 3 \\
\hline
1 & 2 & 0 & 0 & 0 & 0 & 0 & 0
\end{array}
\right]
\begin{array}{l}
\#3 \\ \#2 \\ \\ \#1 \\ \#4 \\ \#5 \\ \\ \
\end{array}
\qquad (2.79)
$$

$$
\begin{array}{cccccccc}
x & y & s_1 & s_2 & s_3 & s_4 & s_5 &
\end{array}
$$

Switching rows 1 and 3 in (2.77) has thrown the identity columns out of whack and we repair this by exchanging the s_1 and s_3 columns,

$$
\left[
\begin{array}{cc|cc|ccc|c}
1 & 0 & 1 & 0 & 0 & 0 & 0 & 2 \\
0 & -1 & 0 & 1 & 0 & 0 & 0 & 0 \\
\hline
-1 & 0 & 0 & 0 & 1 & 0 & 0 & 0 \\
0 & 1 & 0 & 0 & 0 & 1 & 0 & 2 \\
1 & 1 & 0 & 0 & 0 & 0 & 1 & 3 \\
\hline
1 & 2 & 0 & 0 & 0 & 0 & 0 & 0
\end{array}
\right]
\begin{array}{l}
\#3 \\ \#2 \\ \\ \#1 \\ \#4 \\ \#5 \\ \\ \
\end{array}
\qquad (2.80)
$$

$$
\begin{array}{cccccccc}
x & y & s_3 & s_2 & s_1 & s_4 & s_5 &
\end{array}
$$

and renumbering to restore the format of (2.74):

$$
\left[\begin{array}{cc|cc|ccc|c}
1 & 0 & 1 & 0 & 0 & 0 & 0 & 2 \\
0 & -1 & 0 & 1 & 0 & 0 & 0 & 0 \\
\hline
-1 & 0 & 0 & 0 & 1 & 0 & 0 & 0 \\
0 & 1 & 0 & 0 & 0 & 1 & 0 & 2 \\
1 & 1 & 0 & 0 & 0 & 0 & 1 & 3 \\
\hline
1 & 2 & 0 & 0 & 0 & 0 & 0 & 0
\end{array}\right]
\left.\begin{array}{l}
\#1' \\ \#2' \\ \\ \#3' \\ \#4' \\ \#5' \\ \\ \end{array}\right\}\,'
\qquad (2.81)
$$

$$\qquad x \quad y \quad s'_1 \; s'_2 \quad s'_3 \; s'_4 \; s'_5$$

where

$$[\#1' \ \#2' \ \#3' \ \#4' \ \#5'] \equiv [\#3 \ \#2 \ \#1 \ \#4 \ \#5],$$

$$[s'_1 \ s'_2 \ s'_3 \ s'_4 \ s'_5] \equiv [s_3 \ s_2 \ s_1 \ s_4 \ s_5].$$

Tableau (2.81) is our new constraint tableau. We row-reduce:

$$
\left[\begin{array}{cc|cc|ccc|c}
1 & 0 & 1 & 0 & 0 & 0 & 0 & 2 \\
0 & 1 & 0 & -1 & 0 & 0 & 0 & 0 \\
\hline
0 & 0 & 1 & 0 & 1 & 0 & 0 & 2 \\
0 & 0 & 0 & 1 & 0 & 1 & 0 & 2 \\
0 & 0 & -1 & \mathbf{1} & 0 & 0 & 1 & 1 \\
\hline
0 & 0 & -1 & \mathbf{2} & 0 & 0 & 0 & -2
\end{array}\right]
\left.\begin{array}{l}
\#1' \\ \#2' \\ \\ \#3' \\ \#4' \\ \mathbf{\#5'} \\ \\ \end{array}\right\}\,\cdot
\qquad (2.82)
$$

$$\qquad x \quad y \quad s'_1 \; s'_2 \quad s'_3 \; s'_4 \; s'_5$$

The Simplex algorithm has moved us to the corner $x = 2$, $y = 0$ in the polygon and the active slacks are s'_1 and s'_2 (s_3 and s_2 in the original listing (2.76)).

The bold "2" in the s'_2 column of the $\boldsymbol{c}$-row of tableau (2.82) indicates that the *second* slack variable should be deactivated, and the minimal s/t ratio in that column lies in row 5. Thus we exchange rows (2.75) and (2.78) in the most recent constraint tableau (2.81),

$$
\left.\begin{bmatrix}
1 & 0 & | & 1 & 0 & | & 0 & 0 & 0 & | & 2 \\
1 & 1 & | & 0 & 0 & | & 0 & 0 & 1 & | & 3 \\
- & - & - & - & - & - & - & - & - & - & - \\
-1 & 0 & | & 0 & 0 & | & 1 & 0 & 0 & | & 0 \\
0 & 1 & | & 0 & 0 & | & 0 & 1 & 0 & | & 2 \\
0 & -1 & | & 0 & 1 & | & 0 & 0 & 0 & | & 0 \\
- & - & - & - & - & | & - & - & - & - & - \\
1 & 2 & | & 0 & 0 & | & 0 & 0 & 0 & | & 0
\end{bmatrix}\right\}
\begin{matrix}
\#1' \\ \#5' \\ - \\ \#3' \\ \#4' \\ \#2' \\ -
\end{matrix} \;,
\qquad (2.83)
$$

$$
\begin{matrix}
x & y & & s_1' & s_2' & & s_3' & s_4' & s_5'
\end{matrix}
$$

rearrange to repair the column damage,

$$
\left.\begin{bmatrix}
1 & 0 & | & 1 & 0 & | & 0 & 0 & 0 & | & 2 \\
1 & 1 & | & 0 & 1 & | & 0 & 0 & 0 & | & 3 \\
- & - & - & - & - & - & - & - & - & - & - \\
-1 & 0 & | & 0 & 0 & | & 1 & 0 & 0 & | & 0 \\
0 & 1 & | & 0 & 0 & | & 0 & 1 & 0 & | & 2 \\
0 & -1 & | & 0 & 0 & | & 0 & 0 & 1 & | & 0 \\
- & - & - & - & - & | & - & - & - & - & - \\
1 & 2 & | & 0 & 0 & | & 0 & 0 & 0 & | & 0
\end{bmatrix}\right\}
\begin{matrix}
\#1' \\ \#5' \\ - \\ \#3' \\ \#4' \\ \#2' \\ -
\end{matrix} \;,
\qquad (2.84)
$$

$$
\begin{matrix}
x & y & & s_1' & s_5' & & s_3' & s_4' & s_2'
\end{matrix}
$$

and renumber to display our next constraint tableau:

$$
\left.\begin{bmatrix}
1 & 0 & | & 1 & 0 & | & 0 & 0 & 0 & | & 2 \\
1 & 1 & | & 0 & 1 & | & 0 & 0 & 0 & | & 3 \\
- & - & - & - & - & - & - & - & - & - & - \\
-1 & 0 & | & 0 & 0 & | & 1 & 0 & 0 & | & 0 \\
0 & 1 & | & 0 & 0 & | & 0 & 1 & 0 & | & 2 \\
0 & -1 & | & 0 & 0 & | & 0 & 0 & 1 & | & 0 \\
- & - & - & - & - & | & - & - & - & - & - \\
1 & 2 & | & 0 & 0 & | & 0 & 0 & 0 & | & 0
\end{bmatrix}\right\}
\begin{matrix}
\#1'' \\ \#2'' \\ - \\ \#3'' \\ \#4'' \\ \#5'' \\ -
\end{matrix}
$$

$$
\begin{matrix}
x & y & & s_1'' & s_2'' & & s_3'' & s_4'' & s_5''
\end{matrix}
$$

$$
[\#1''\ \#2''\ \#3''\ \#4''\ \#5''] \equiv [\#1'\ \#5'\ \#3'\ \#4'\ \#2''],
$$

$$
[s_1''\ s_2''\ s_3''\ s_4''\ s_5''] \equiv [s_1'\ s_5'\ s_3'\ s_4'\ s_2'].
$$

Apply row reduction:

$$
\left[\begin{array}{cc|cc|ccc|c}
1 & 0 & 1 & 0 & 0 & 0 & 0 & 2 \\
0 & 1 & -1 & 1 & 0 & 0 & 0 & 1 \\
\hline
0 & 0 & 1 & 0 & 1 & 0 & 0 & 2 \\
0 & 0 & 1 & -1 & 0 & 1 & 0 & 1 \\
0 & 0 & -1 & 1 & 0 & 0 & 1 & 1 \\
\hline
0 & 0 & 1 & -2 & 0 & 0 & 0 & -4
\end{array}\right]
\begin{array}{l}
\left.\begin{array}{l}\#1'' \\ \#2''\end{array}\right) \\[4pt]
\left.\begin{array}{l}\#3'' \\ \#4'' \\ \#5''\end{array}\right\} \\[8pt]
\end{array}
\qquad (2.85)
$$

$$\underbrace{}_{\begin{array}{cc} x & y\end{array}}\ \underbrace{}_{\begin{array}{cc}s_1'' & s_2''\end{array}}\ \underbrace{}_{\begin{array}{ccc}s_3'' & s_4'' & s_5''\end{array}}$$

Tableau (2.86) dictates that we deactivate the first slack variable and activate the fourth; we switch rows 1 and 4 in the most recent constraint tableau (2.85):

$$
\left[\begin{array}{cc|cc|ccc|c}
0 & 1 & 0 & 0 & 0 & 1 & 0 & 2 \\
1 & 1 & 0 & 1 & 0 & 0 & 0 & 3 \\
\hline
-1 & 0 & 0 & 0 & 1 & 0 & 0 & 0 \\
1 & 0 & 1 & 0 & 0 & 0 & 0 & 2 \\
0 & -1 & 0 & 0 & 0 & 0 & 1 & 0 \\
\hline
1 & 2 & 0 & 0 & 0 & 0 & 0 & 0
\end{array}\right]
\begin{array}{l}
\left.\begin{array}{l}\#4'' \\ \#2''\end{array}\right) \\[4pt]
\left.\begin{array}{l}\#3'' \\ \#1'' \\ \#5''\end{array}\right\} \\[8pt]
\end{array}
\qquad (2.86)
$$

$$\underbrace{}_{\begin{array}{cc} x & y\end{array}}\ \underbrace{}_{\begin{array}{cc}s_1'' & s_2''\end{array}}\ \underbrace{}_{\begin{array}{ccc}s_3'' & s_4'' & s_5''\end{array}}$$

sort the columns,

$$
\left[\begin{array}{cc|cc|ccc|c}
0 & 1 & 1 & 0 & 0 & 0 & 0 & 2 \\
1 & 1 & 0 & 1 & 0 & 0 & 0 & 3 \\
\hline
-1 & 0 & 0 & 0 & 1 & 0 & 0 & 0 \\
1 & 0 & 0 & 0 & 0 & 1 & 0 & 2 \\
0 & -1 & 0 & 0 & 0 & 0 & 1 & 0 \\
\hline
1 & 2 & 0 & 0 & 0 & 0 & 0 & 0
\end{array}\right]
\begin{array}{l}
\left.\begin{array}{l}\#4'' \\ \#2''\end{array}\right) \\[4pt]
\left.\begin{array}{l}\#3'' \\ \#1'' \\ \#5''\end{array}\right\} \\[8pt]
\end{array}
\qquad (2.87)
$$

$$\underbrace{}_{\begin{array}{cc} x & y\end{array}}\ \underbrace{}_{\begin{array}{cc}s_4'' & s_2''\end{array}}\ \underbrace{}_{\begin{array}{ccc}s_3'' & s_1'' & s_5''\end{array}}$$

renumber to display the next constraint tableau,

$$
\left[
\begin{array}{cc|cc|ccc|c}
0 & 1 & 1 & 0 & 0 & 0 & 0 & 2 \\
1 & 1 & 0 & 1 & 0 & 0 & 0 & 3 \\
\hline
-1 & 0 & 0 & 0 & 1 & 0 & 0 & 0 \\
1 & 0 & 0 & 0 & 0 & 1 & 0 & 2 \\
0 & -1 & 0 & 0 & 0 & 0 & 1 & 0 \\
\hline
1 & 2 & 0 & 0 & 0 & 0 & 0 & 0
\end{array}
\right]
\quad
\left.
\begin{array}{l}
\#1''' \\
\#2''' \\
\\
\#3''' \\
\#4''' \\
\#5''' \\
\\
\\
\end{array}
\right\}
\qquad (2.88)
$$

$$
\begin{array}{ccccccc}
x & y & s_1''' & s_2''' & & s_3''' & s_4''' & s_5'''
\end{array}
$$

$$
\left[\#1'''\ \#2'''\ \#3'''\ \#4'''\ \#5'''\right] \equiv \left[\#4''\ \#2''\ \#3''\ \#1''\ \#5''\right]
$$

$$
\left[s_1'''\ s_2'''\ s_3'''\ s_4'''\ s_5'''\right] \equiv \left[s_4''\ s_2''\ s_3''\ s_1''\ s_5''\right],
$$

and row-reduce:

$$
\left[
\begin{array}{cc|cc|ccc|c}
0 & 1 & -1 & 1 & 0 & 0 & 0 & 1 \\
1 & 1 & 1 & 0 & 0 & 0 & 0 & 2 \\
\hline
0 & 0 & -1 & 1 & 1 & 0 & 0 & 1 \\
0 & 0 & 1 & -1 & 0 & 1 & 0 & 1 \\
0 & 0 & 1 & 0 & 0 & 0 & 1 & 2 \\
\hline
0 & 0 & -1 & -1 & 0 & 0 & 0 & -5
\end{array}
\right]
\quad
\left.
\begin{array}{l}
\#1''' \\
\#2''' \\
\\
\#3''' \\
\#4''' \\
\#5''' \\
\\
\end{array}
\right\}
\qquad (2.89)
$$

$$
\begin{array}{cccccccc}
x & y & s_1''' & s_2''' & s_3''' & s_4''' & s_5'''
\end{array}
$$

According to the c-row we're done! The maximum of the objective function is $(+)5$, occurring at the corner $x = 1$, $y = 2$.

2.8 Streamlining the Simplex Algorithm

The procedure in Sects. 2.6 and 2.7 was designed for clarity, not for efficiency. Now we turn to streamlining the computations. To keep the notation manageable, we'll demonstrate with the two-dimensional Example 2.2 of Sect. 2.7. (The generalization to higher dimensions will be apparent.)

Recall each stage of the procedure begins with some arrangement of the *constraint* tableau, displaying the constraints (active ones first) and the objective function

$$
\begin{bmatrix}
n_1 \to & | & & | & & | & d_1 \\
n_2 \to & | & I_{2\times 2} & | & 0_{2\times 3} & | & d_2 \\
- - - & | & - - - & | & - - - & | & - \\
n_3 \to & | & & | & & | & d_3 \\
n_4 \to & | & 0_{3\times 2} & | & I_{3\times 3} & | & d_4 \\
n_5 \to & | & & | & & | & d_5 \\
- - - & | & - - - & | & - - - & | & - \\
c \to & | & 0 \quad 0 & | & 0 \quad 0 \quad 0 & | & 0
\end{bmatrix}
\begin{matrix}
\#1 \\ \#2 \\ - \\ \#3 \\ \#4 \\ \#5 \\ - \\
\end{matrix}
\qquad (2.90)
$$

$$
\underbrace{}_{x \quad y} \quad \underbrace{}_{s_1 \quad s_2} \quad \underbrace{}_{s_3 \ s_4 \ s_5}
$$

and ends with a *row-reduced* tableau—a blueprint that neatly sorts the coordinate, slack variable, and objective function values and the data necessary for selecting the next corner for analysis:

$$
\begin{bmatrix}
I_{2\times 2} & | & N^{-1} & | & 0_{2\times 3} & | & \begin{matrix} x \\ y \end{matrix} \\
- - - & | & - - - & | & & | & - - \\
0 \quad 0 & | & t_3^{(1)} \ t_3^{(2)} & | & & | & s_3 \\
0 \quad 0 & | & t_4^{(1)} \ t_4^{(2)} & | & I_{3\times 3} & | & s_4 \\
0 \quad 0 & | & t_5^{(1)} \ t_5^{(2)} & | & & | & s_5 \\
- - - & | & - - - & | & - - - & | & - - \\
0 \quad 0 & | & -cN^{-1} & | & 0 \quad 0 \quad 0 & | & -c \cdot R
\end{bmatrix}
\qquad (2.91)
$$

$$
\underbrace{}_{x \quad y} \quad \underbrace{}_{s_1 \quad s_2} \quad \underbrace{}_{s_3 \ s_4 \ s_5}
$$

For instance, pivoting was employed to row-reduce the initial constraint tableau for Example 2.2 (2.77, Sect. 2.7) to

$$
\begin{bmatrix}
1 & 0 & | & -1 & 0 & | & 0 & 0 & 0 & | & 0 \\
0 & 1 & | & 0 & -1 & | & 0 & 0 & 0 & | & 0 \\
- & - & | & - & - & | & - & - & - & | & - \\
0 & 0 & | & 1 & 0 & | & 1 & 0 & 0 & | & 2 \\
0 & 0 & | & 0 & 1 & | & 0 & 1 & 0 & | & 2 \\
0 & 0 & | & 1 & 1 & | & 0 & 0 & 1 & | & 3 \\
- & - & | & - & - & | & - & - & - & | & - \\
0 & 0 & | & 1 & 2 & | & 0 & 0 & 0 & | & 0
\end{bmatrix}
\begin{matrix}
\#1 \\ \#2 \\ - \\ \#3 \\ \#4 \\ \#5 \\ - \\
\end{matrix}
\quad . \, (2.78, \text{Sect. } 2.7)
$$

$$
\underbrace{}_{x \quad y} \quad \underbrace{}_{s_1 \quad s_2} \quad \underbrace{}_{s_3 \ s_4 \ s_5}
$$

This tableau reminded us that s_1 and s_2 were the active slack variables, and predicted that the objective function would be increased if s_1 were deactivated at the cost of activating s_3.

We are going to focus on the constraint submatrix, and ignore the c-row, for a while. Notice in the row-reduced tableau (2.91) that this submatrix, which is the augmented matrix for the linear system of constraint equations, contains all five columns of the 5-by-5 identity matrix. Therefore we can apply Sect. 2.3's *basic solution* concept to it; it possesses a solution wherein all the variables associated with the cluttered columns are zero, and the values of the other variables—the "basic variables"—are displayed in the final column. But the labels in the final column of (2.91) say precisely the same thing; x, y, s_3, s_4, and s_5 are associated with the identity columns. And the columns for the active slack variables—which are zero—are cluttered.

> *The solution displayed in a row-reduced tableau coincides with the "basic solution" for its constraint submatrix.*[21]

Continuing to scrutinize Example 2.2: after the rearranging/pivoting directives indicated by tableau (2.78, Sect. 2.7) were applied to the first constraint tableau (2.77, Sect. 2.7), in the subsequent row-reduced tableau *the column for* s_1 *(now s_3') became an identity column while the column for* s_3 *(now s_1') became cluttered:*

$$
\left[\begin{array}{cc|cc|ccc|c}
1 & 0 & 1 & 0 & 0 & 0 & 0 & 2 \\
0 & 1 & 0 & -1 & 0 & 0 & 0 & 0 \\
\hline
0 & 0 & 1 & 0 & 1 & 0 & 0 & 2 \\
0 & 0 & 0 & 1 & 0 & 1 & 0 & 2 \\
0 & 0 & -1 & 1 & 0 & 0 & 1 & 1 \\
\hline
0 & 0 & -1 & 2 & 0 & 0 & 0 & -2
\end{array}\right]
\begin{array}{l}
\#1' \\ \#2' \\ \\ \#3' \\ \#4' \\ \#5' \\ \\ \end{array}
\quad . \;(2.82,\ \text{Sect. }2.7)
$$

$$
\begin{array}{cccccc}
x & y & s_1' & s_2' & s_3' & s_4' & s_5'
\end{array}
$$

$$
[\#1'\ \#2'\ \#3'\ \#4'\ \#5'] \equiv [\#3\ \#2\ \#1\ \#4\ \#5],
$$

$$
[s_1'\ s_2'\ s_3'\ s_4'\ s_5'] \equiv [s_3\ s_2\ s_1\ s_4\ s_5],
$$

All of the other columns retained their cluttered/uncluttered character.

If we didn't care about the placement of the columns, we could have accomplished these column conversions *in one step*: pivot the first tableau (2.78, Sect. 2.7) *in place* on the s/t competition winner—that is, on the 1 in the third row, third column. This results in

---

[21] This statement does not hold for the *constraint* tableaus, despite the fact that they, too, contain the identity columns. It fails, for example, for the second constraint tableau in Example 2.2 ((2.6), Sect. 2.7), where the x column is cluttered but $x = 2$ at the corner. A little thought reveals that the x and y columns are always cluttered in the constraint tableaux, but $x = y = 0$ only if the corner is at the origin, as in ((2.4), Sect. 2.7).

$$
\begin{bmatrix}
1 & 0 & | & 0 & 0 & | & 1 & 0 & 0 & | & 2 \\
0 & 1 & | & 0 & -1 & | & 0 & 0 & 0 & | & 0 \\
- & - & | & - & - & | & - & - & - & | & - \\
0 & 0 & | & 1 & 0 & | & 1 & 0 & 0 & | & 2 \\
0 & 0 & | & 0 & 1 & | & 0 & 1 & 0 & | & 2 \\
0 & 0 & | & 0 & 1 & | & -1 & 0 & 1 & | & 1 \\
- & - & | & - & - & | & - & - & - & | & - \\
0 & 0 & | & 0 & 2 & | & -1 & 0 & 0 & | & -2
\end{bmatrix}
\begin{matrix}
\#1 \\ \#2 \\ \\ \#3 \\ \#4 \\ \#5 \\ \\ \\
\end{matrix}
\quad \cdot \qquad (2.92)
$$

$$x \quad y \qquad s_1 \quad s_2 \qquad\quad s_3 \quad s_4 \quad s_5$$

Now s_1 anchors an identity column, the column of s_3 is cluttered, and the other retain their character. *If we restore the column order in (2.82, Sect. 2.7), it possesses the same collection of uncluttered columns as (2.92); thus, since these two tableaux represent equivalent linear systems, they have the same basic solution*—that is, the tableaux display the same solution information. And with (2.92) we have leapfrogged the row/column exchanges and *double* pivoting that burdened the computation of (2.82, Sect. 2.7)!

That's encouraging. But what about the *final* row in (2.92); is it reliable for providing the objective function $c \cdot R$ value, and dictating the next constraint to be deactivated?[22]

Remember that the c-row does not represent an equation; the significance of its disclosures was cited merely as a "bonus" in Sect. 2.6. But we are more accustomed to interpreting matrix rows as equations. Thus, to enfranchise the row as an equation, first we (temporarily) create a new symbol for the objective function:

$$
\sigma \equiv c \cdot R = \begin{cases} c_1 x + c_2 y & \text{in 2 dimensions} \\ c_1 x + c_2 y + c_3 z & \text{in 3 dimensions} \end{cases} . \qquad (2.93)
$$

Bear in mind that $\sigma = c_1 x + c_2 y$ is a function of x and y (in two dimensions); if x and y were expressed in terms of some other variables, say

$$x = u + v, \ y = u - v + 3, \qquad (2.94)$$

then σ would satisfy the equation[23]

$$\sigma = c_1 x + c_2 y = (c_1 + c_2)u + (c_1 - c_2)v + 3c_2 \equiv \tilde{c}_1 x + \tilde{c}_2 y + 3c_2. \qquad (2.95)$$

Now let us go back to the template (2.90) for the starting constraint tableau and (temporarily) alter it so that the c-row states the *genuine* Eq. (2.93) in the form $c_1 x + c_2 y - \sigma = 0$. This requires inserting a new column for σ:

---

[22] It certainly appears so, from this one experiment; (2.3) and ((2.9), Sect. 2.7) coincide if the original row/column numbering is restored.

[23] (Because $\sigma = \sigma(x, y) = \sigma(x(u, v), y(u, v)) = c_1 x(u, v) + c_2 y(u, v)$
$= c_1(u + v) + c_2(u - v + 3) = (c_1 + c_2)u + (c_1 - c_2)v + 3c_2$).

$$
\begin{bmatrix}
n_1 \rightarrow & & & & & & 0 & & d_1 \\
n_2 \rightarrow & & I_{2\times2} & & 0_{2\times3} & & \vdots & & d_2 \\
\hline
n_3 \rightarrow & & & & & & 0 & & d_3 \\
n_4 \rightarrow & & 0_{3\times2} & & I_{3\times3} & & \vdots & & d_4 \\
n_5 \rightarrow & & & & & & 0 & & d_5 \\
\hline
c_1 \;\; c_2 & & 0 \;\; 0 & & 0 \;\; 0 \;\; 0 & & -1 & & 0
\end{bmatrix}.
$$

$$
\underbrace{}_{x \;\; y} \quad \underbrace{}_{s_1 \;\; s_2} \quad \underbrace{}_{s_3 \;\; s_4 \;\; s_5 \;\; \sigma}
\qquad\qquad (2.96)
$$

We ask: if we select any nonzero entry from, say, the first column in (2.96)'s constraint submatrix and pivoted on it, what would be the result? Of course the first column would become an identity column and the corresponding identity column would become cluttered. But more significantly, according to "Equation Content Immunity" in Sect. 2.3 the new c-row would simply reiterate $c_1 x + c_2 y - \sigma = 0$ with x replaced by an equivalent formula in terms of the other variables:

$$
x = \alpha y + \beta s_1 + \cdots + \mu s_5 + \rho
$$

$$
c_1 x + c_2 y - \sigma = c_1(\alpha y + \beta s_1 + \cdots + \mu s_5 + \rho) + c_2 y - \sigma = 0
$$

rearranged as

$$
\underbrace{\tilde{c}_1}_{=0} x + \tilde{c}_2 y + \tilde{c}_3 s_1 + \tilde{c}_4 s_2 + \cdots + \tilde{c}_7 s_5 - \sigma = \theta
$$

$$
(2.97)
$$

(compare Eq. (2.18, Sect. 2.3). For instance if Example 2.2's initial constraint tableau (2.77, Sect. 2.7) is fitted with a σ-column

$$
\begin{bmatrix}
-1 & 0 & | & 1 & 0 & | & 0 & 0 & 0 & 0 & | & 0 \\
0 & -1 & | & 0 & 1 & | & 0 & 0 & 0 & 0 & | & 0 \\
\hline
1 & 0 & | & 0 & 0 & | & 1 & 0 & 0 & 0 & | & 2 \\
0 & 1 & | & 0 & 0 & | & 0 & 1 & 0 & 0 & | & 2 \\
1 & 1 & | & 0 & 0 & | & 0 & 0 & 1 & 0 & | & 3 \\
\hline
1 & 2 & | & 0 & 0 & | & 0 & 0 & 0 & -1 & | & 0
\end{bmatrix}
\begin{matrix}
\#1 \\ \#2 \\ \\ \#3 \\ \#4 \\ \#5 \\ \\ \end{matrix}
',
\qquad (2.98)
$$

$$
\underbrace{}_{x \;\; y} \quad \underbrace{}_{s_1 \;\; s_2} \quad \underbrace{}_{s_3 \;\; s_4 \;\; s_5 \;\; \sigma}
$$

and we pivot on, say, the $(3, 1)$ entry "1" *in place*, we get

$$
\left[
\begin{array}{cc|cc|cccc|c}
0 & 0 & 1 & 0 & 1 & 0 & 0 & \textbf{0} & 2 \\
0 & -1 & 0 & 1 & 0 & 0 & 0 & \textbf{0} & 0 \\
\hline
1 & 0 & 0 & 0 & 1 & 0 & 0 & \textbf{0} & 2 \\
0 & 1 & 0 & 0 & 0 & 1 & 0 & \textbf{0} & 2 \\
0 & 1 & 0 & 0 & -1 & 0 & 1 & \textbf{0} & 1 \\
\hline
0 & 2 & 0 & 0 & -1 & 0 & 0 & \textbf{-1} & \textbf{-2}
\end{array}
\right]
\begin{array}{l}
\#1 \\ \#2 \\ \\ \#3 \\ \#4 \\ \#5 \\ \\ \\
\end{array}
\qquad (2.99)
$$

$$
\underset{x\quad y\quad\quad s_1\quad s_2\quad\quad s_3\quad s_4\quad s_5\quad \sigma}{}
$$

(where "-2" is the "θ" in Eq. (2.97)). Now the relation (2.4) ($\sigma = c_1 x + c_2 y$) still holds, but the $\boldsymbol{c}$-row in (2.99) states that it is equivalent to the relation[24]

$$
2y - 1s_3 - \sigma = -2, \quad \text{or} \quad \sigma \equiv \underbrace{\boldsymbol{c}\cdot\boldsymbol{R}}_{c_1 x + c_2 y} = 2y - 1s_3 - (-2). \qquad (2.100)
$$

This enables us to discern the significance of the "-2", or the "θ", in the $\boldsymbol{c}$-row of (2.99). When we apply the "basic solution" interpretation to the constraint submatrix in (2.99), we set the variables y and s_3 equal to zero. Thus Eq. (2.100) states that the "-2" is the value of $-\boldsymbol{c}\cdot\boldsymbol{R}$ when the active variables are set equal to zero. *The lower right entry of the pivoted tableau* (2.99) *is, indeed, the negative of the value of the objective function at the current corner.*

By the same argument, any further pivoting in the tableau will produce a $\boldsymbol{c}$-row of the form

$$
\underset{x\quad y\quad s_1\quad s_2\quad s_3\quad s_4\quad s_5\quad \sigma}{\left[\tilde{c}_1 \quad \tilde{c}_2 \quad \tilde{c}_3 \quad \tilde{c}_4 \quad \tilde{c}_5 \quad \tilde{c}_6 \quad \tilde{c}_7 \quad -1 \ \vdots \ \theta\right]} \quad (\boldsymbol{c}\text{-row}) \qquad (2.101)
$$

with all but two of the $\tilde{c}_k$'s (say, $\tilde{c}_\alpha$ and $\tilde{c}_\beta$) equal to zero (i.e., two cluttered columns). Equation (2.101) has the form

$$
\tilde{c}_\alpha s_\gamma + \tilde{c}_\beta s_\delta - \boldsymbol{c}\cdot\boldsymbol{R} \equiv \tilde{c}_\alpha s_\gamma + \tilde{c}_\beta s_\delta - \sigma = \theta, \qquad (2.102)
$$

identifying θ as the value of $-\boldsymbol{c}\cdot\boldsymbol{R}$ when s_β, s_δ become the new active slack variables. *The role of the final entry in the altered row-reduced matrices*

---

[24] (And if we had pivoted on a different entry in the constraint submatrix we would have obtained another (still equivalent) expression of formula (2.93)).

$$\left[\begin{array}{cc|cc|cccc|c}
 & & & & & & 0 & & x \\
\mathbf{I}_{2\times2} & & \mathbf{N}^{-1} & & \mathbf{0}_{2\times3} & & \vdots & & y \\
\hline
0 & 0 & t_3^{(1)} & t_3^{(2)} & & & & 0 & s_3 \\
0 & 0 & t_4^{(1)} & t_4^{(2)} & \mathbf{I}_{3\times3} & & \vdots & & s_4 \\
0 & 0 & t_5^{(1)} & t_5^{(2)} & & & & 0 & s_5 \\
\hline
0 & 0 & -\mathbf{c}\mathbf{N}^{-1} & & 0 & 0 & 0 & -1 & -\mathbf{c}\cdot\mathbf{R}
\end{array}\right] \qquad (2.103)$$

$$\underbrace{}_{x}\ \underbrace{}_{y}\quad \underbrace{}_{s_1}\ \underbrace{}_{s_2}\quad \underbrace{}_{s_3}\ \underbrace{}_{s_4}\ \underbrace{}_{s_5}\ \underbrace{}_{\sigma}$$

is not merely a serendipitous by product of the diamond analogy of Sect. 2.4; it is confirmed by matrix analysis.

Expounding further on the $\mathbf{c}$-row, reconsider the significance of

$$\tilde{c}_\alpha s_\gamma + \tilde{c}_\beta s_\delta - \mathbf{c}\cdot\mathbf{R} \equiv \tilde{c}_\alpha s_\gamma + \tilde{c}_\beta s_\delta - \sigma = \theta, \quad (2.102 \text{ repeated})$$

when one of the coefficients—say $\tilde{c}_\alpha$—is positive. If we are at a corner where s_γ and s_δ are zero, then (2.102) implies that if we increase s_γ, then $\sigma = \mathbf{c}\cdot\mathbf{R}$ will increase to balance the equality (since θ is independent of the variables). This generalizes the argument in Sect. 2.4: *deactivating an active slack variable anchoring a positive entry in its $\mathbf{c}$-row will increase the objective function.*

Finally, Sect. 2.4's diamond-analogy argument concerning the "s/t ratio" competition is similarly validated by matrix analysis: suppose we have a row-reduced version of the tableau such as (2.99):

$$\left[\begin{array}{cc|cc|cccc|c}
0 & 0 & 1 & 0 & 1 & 0 & 0 & 0 & 2 \\
0 & -1 & 0 & 1 & 0 & 0 & 0 & 0 & 0 \\
\hline
1 & 0 & 0 & 0 & 1 & 0 & 0 & 0 & 2 \\
0 & 1 & 0 & 0 & 0 & 1 & 0 & 0 & 2 \\
0 & 1 & 0 & 0 & -1 & 0 & 1 & 0 & 1 \\
\hline
0 & 2 & 0 & 0 & -1 & 0 & 0 & -1 & -2
\end{array}\right]. \quad (2.99 \text{ repeated})$$

$$x \quad y \qquad s_1 \quad s_2 \qquad s_3 \quad s_4 \quad s_5 \quad \sigma$$

The (zero-valued) variables associated with the cluttered columns are y and s_3, and the positive 2 in the second column of the $\mathbf{c}$-row dictates that we should increase y. With t denoting a positive entry in the second column and s denoting the entry in the final column of the same row, we see that the row's equation can accommodate an increase of y on the left by decreasing s on the right by ty. So y can be increased by as much as s/t without violating the nonnegative condition on the slack variables, for that particular row.[25] Therefore as y increases no nonnegativity violation will occur in any row until y reaches the lowest s/t value. The validity of the s/t competition is established in full generality.

> To sum up: the strategies of the Simplex one-step update of a row-reduced tableau, Eq. (2.92), have been validated by matrix analysis, without appealing to the geometric analogy of Sect. 2.4. We have shown that if any row-reduced Simplex tableau, possessing a positive entry in its **c**-row and defining a legitimate corner per the basic-solution interpretation, is pivoted on a positive entry lying in the same column and minimizing the s/t ratios, the resulting matrix will be another Simplex tableau defining another legitimate corner (and suitable for further pivoting).

That completes mathematical validation of the streamlined update.

> *Note that the σ-column that we introduced never changed; so we don't need it anymore.* We'll drop it from now on and forget that it ever existed.

We shall demonstrate the Streamlined Simplex Algorithm for Example 2.2, Sect. 2.7. As you scan these results, just focus on the highlighted parameters, noting the pivoting strategy. We begin, as in the tried and true procedure of Sect. 2.6, by pivoting to install an identity in the upper left corner of the first constraint tableau:

$$
\begin{bmatrix}
-1 & 0 & | & 1 & 0 & | & 0 & 0 & 0 & | & 0 \\
0 & -1 & | & 0 & 1 & | & 0 & 0 & 0 & | & 0 \\
- & - & | & - & - & | & - & - & - & | & - \\
1 & 0 & | & 0 & 0 & | & 1 & 0 & 0 & | & 2 \\
0 & 1 & | & 0 & 0 & | & 0 & 1 & 0 & | & 2 \\
1 & 1 & | & 0 & 0 & | & 0 & 0 & 1 & | & 3 \\
- & - & | & - & - & | & - & - & - & | & - \\
1 & 2 & | & 0 & 0 & | & 0 & 0 & 0 & | & 0
\end{bmatrix}
\begin{matrix}
\#1 \\
\#2 \\
- \\
\#3 \\
\#4 \\
\#5 \\
- \\
\end{matrix}
\quad . (2.77, \text{Sect. } 2.7)
$$

$$\qquad x \quad\ y \qquad s_1 \quad s_2 \qquad s_3 \quad s_4 \quad s_5$$

$$\Downarrow$$

---

[25] (And zero or negative values of t will never result in a violation).

$$
\begin{bmatrix}
1 & 0 & | & -1 & 0 & | & 0 & 0 & 0 & | & 0 \\
0 & 1 & | & 0 & -1 & | & 0 & 0 & 0 & | & 0 \\
- & - & | & - & - & | & - & - & - & | & - \\
0 & 0 & | & 1 & 0 & | & 1 & 0 & 0 & | & 2 \\
0 & 0 & | & 0 & 1 & | & 0 & 1 & 0 & | & 2 \\
0 & 0 & | & 1 & 1 & | & 0 & 0 & 1 & | & 3 \\
- & - & | & - & - & | & - & - & - & | & - \\
0 & 0 & | & 1 & 2 & | & 0 & 0 & 0 & | & 0
\end{bmatrix}
\begin{matrix}
\#1 \\ \#2 \\ \\ \#3 \\ \#4 \\ \#5 \\ \\ \end{matrix}
. \text{(2.78, Sect. 2.7)}
$$

$$
\begin{matrix} x & y & & s_1 & s_2 & & s_3 & s_4 & s_5 \end{matrix}
$$

We choose to deactivate s_1 (s_2 would also be acceptable). The location of the winner "1" of the s/t competition tells us to activate the third constraint. Now we simply pivot on the s/t winner, the "1", *in place*:

$$
\begin{bmatrix}
1 & 0 & | & 0 & 0 & | & 1 & 0 & 0 & | & 2 \\
0 & 1 & | & 0 & -1 & | & 0 & 0 & 0 & | & 0 \\
- & - & | & - & - & | & - & - & - & | & - \\
0 & 0 & | & 1 & 0 & | & 1 & 0 & 0 & | & 2 \\
0 & 0 & | & 0 & 1 & | & 0 & 1 & 0 & | & 2 \\
0 & 0 & | & 0 & 1 & | & -1 & 0 & 1 & | & 1 \\
- & - & | & - & - & | & - & - & - & | & - \\
0 & 0 & | & 0 & 2 & | & -1 & 0 & 0 & | & -2
\end{bmatrix}
\begin{matrix}
\#1 \\ \#2 \\ \\ \#3 \\ \#4 \\ \#5 \\ \\ \end{matrix}
. \text{(2.92 repeated)}
$$

$$
\begin{matrix} x & y & & s_1 & s_2 & & s_3 & s_4 & s_5 \end{matrix}
$$

We conduct the s/t competition in the s_2 column; and the winner is the "1" in row 5. Pivoting again, we obtain

$$
\begin{bmatrix}
1 & 0 & | & 0 & 0 & | & 1 & 0 & 0 & | & 2 \\
0 & 1 & | & 0 & 0 & | & -1 & 0 & 1 & | & 1 \\
- & - & | & - & - & | & - & - & - & | & - \\
0 & 0 & | & 1 & 0 & | & 1 & 0 & 0 & | & 2 \\
0 & 0 & | & 0 & 0 & | & 1 & 1 & -1 & | & 1 \\
0 & 0 & | & 0 & 1 & | & -1 & 0 & 1 & | & 1 \\
- & - & | & - & - & | & - & - & - & | & - \\
0 & 0 & | & 0 & 0 & | & 1 & 0 & -2 & | & -4
\end{bmatrix}
\begin{matrix}
\#1 \\ \#2 \\ \\ \#3 \\ \#4 \\ \#5 \\ \\ \end{matrix}
\quad (2.104)
$$

$$
\begin{matrix} x & y & & s_1 & s_2 & & s_3 & s_4 & s_5 \end{matrix}
$$

The s/t winner is the "1" in the (4, 5) entry; we pivot on it:

$$
\begin{bmatrix}
1 & 0 & | & 0 & 0 & | & 0 & -1 & 1 & | & 1 \\
0 & 1 & | & 0 & 0 & | & 0 & 1 & 0 & | & 2 \\
- & - & - & - & - & - & - & - & - & - & - \\
0 & 0 & | & 1 & 0 & | & 0 & -1 & 1 & | & 1 \\
0 & 0 & | & 0 & 0 & | & 1 & 1 & -1 & | & 1 \\
0 & 0 & | & 0 & 1 & | & 0 & 1 & 0 & | & 2 \\
- & - & - & - & - & | & - & - & - & - & - \\
0 & 0 & | & 0 & 0 & | & 0 & -1 & -1 & | & -5
\end{bmatrix}
\qquad (2.105)
$$

$$
\begin{array}{ccccccc}
x & y & s_1 & s_2 & s_3 & s_4 & s_5
\end{array}
$$

We're done! The maximum objective, $(+)5$, occurs at the corner $x = 1, y = 2$. That agrees with the calculations in Sect. 2.7.

The streamlined procedure took us from the starting tableau (2.77, Sect. 2.7) to the final tableau (2.105) with no row/column swaps and 5 pivots. Sect. 2.7's computation required 8 pivots and 3 row/column exchanges. That's an impressive improvement: roughly half the pivoting, no bookkeeping.

Question 12. The first step in the streamlined implementation discussed here begins exactly the same as the old tried-and-true formulation: shift the active constraint rows to the top of the initial constraint tableau, then pivot to install an identity in the upper left hand corner to obtain the initial row-reduced tableau, then finish up by proceeding with the streamlined updating program. Is there a way we can simply initiate the streamlined program on the initial constraint tableau without shifting the active constraint rows to the top?

Answer. Yes. The basic motive of Sect. 2.6's shift-to-the-top-and-pivot "warmup" was to achieve the visual simplicity of the row-reduced tableau (2.91). But in the course of pruning Example 2.2 we have seen how pivoting *in place* will accomplish the same data processing (including the *c*-row), if we can put up with the shuffling of the locations of the results. Moreover, the streamlined Simplex update doesn't care about the data *locations*; it simply pivots on the s/t competition winner in a column having a positive entry in its *c*-row.

Therefore if we know of two constraint rows that correspond to a legitimate corner, we can pivot in their first two columns (establishing identity columns there) and proceed directly with streamlined Simplex updating.

Let's demonstrate with Example 2.2. From the constraint list

$$
\begin{cases}
\#1 & -x + s_1 = 0 & active\,(s_1 = 0) \\
\#2 & -y + s_2 = 0 & active\,(s_2 = 0) \\
\#3 & x + s_3 = 2 & inactive\,(s_3 = 2) \quad (2.40,\ \text{Sect. 2.6)}) \\
\#4 & x + s_4 = 2 & inactive\,(s_4 = 2) \\
\#5 & x + y + s_5 = 3 & inactive\,(s_5 = 3).
\end{cases}
$$

we created the initial constraint tableau

$$
\left[\begin{array}{cc|cc|ccc|c}
-1 & 0 & 1 & 0 & 0 & 0 & 0 & 0 \\
0 & -1 & 0 & 1 & 0 & 0 & 0 & 0 \\
- & - & - & - & - & - & - & - \\
1 & 0 & 0 & 0 & 1 & 0 & 0 & 2 \\
0 & 1 & 0 & 0 & 0 & 1 & 0 & 2 \\
1 & 1 & 0 & 0 & 0 & 0 & 1 & 3 \\
- & - & - & - & & - & - & - \\
1 & 2 & 0 & 0 & 0 & 0 & 0 & 0
\end{array}\right]
\begin{array}{l}
\#1 \\ \#2 \\ - \\ \#3 \\ \#4 \\ \#5 \\ -
\end{array}
$$

$$\underbrace{}_{x\ \ y}\ \underbrace{}_{s_1\ s_2}\ \underbrace{}_{s_3\ s_4\ s_5}$$

$$. \quad (2.77,\ \text{Sect. } 2.7)$$

We had observed that the first two rows specify the intersection point $x = y = 0$, which is shown to be a legitimate corner in Fig. 2.14. However, the figure shows that $x = 2, y = 0$ is also a legitimate corner, corresponding to the constraints in rows #3 and 2; so for the sake of illustration let's start from this corner, but forego the exchange of rows 1 and 3).

By pivoting on the 1 and the -1 in (2.77, Sect. 2.7) we'll convert their columns to identity columns:

$$
\left[\begin{array}{cc|cc|ccc|c}
0 & 0 & 1 & 0 & 1 & 0 & 0 & 2 \\
0 & 1 & 0 & -1 & 0 & 0 & 0 & 0 \\
- & - & - & - & - & - & - & - \\
1 & 0 & 0 & 0 & 1 & 0 & 0 & 2 \\
0 & 0 & 0 & 1 & 0 & 1 & 0 & 2 \\
0 & 0 & 0 & 1 & -1 & 0 & 1 & 1 \\
- & - & - & - & & - & - & - \\
0 & 0 & 0 & 2 & -1 & 0 & 0 & -2
\end{array}\right]
\begin{array}{l}
\#1 \\ \#2 \\ - \\ \#3 \\ \#4 \\ \#5 \\ -
\end{array}
\qquad (2.106)
$$

$$\underbrace{}_{x\ \ y}\ \underbrace{}_{s_1\ s_2}\ \underbrace{}_{s_3\ s_4\ s_5}$$

Now we immediately apply streamlined Simplex updates, pivoting as indicated by the highlighting:

$$
\left[\begin{array}{cc|cc|ccc|c}
0 & 0 & 1 & 0 & 1 & 0 & 0 & 2 \\
0 & 1 & 0 & 0 & -1 & 0 & 1 & 1 \\
- & - & - & - & - & - & - & - \\
1 & 0 & 0 & 0 & 1 & 0 & 0 & 2 \\
0 & 0 & 0 & 0 & 1 & 1 & -1 & 1 \\
0 & 0 & 0 & 1 & -1 & 0 & 1 & 1 \\
- & - & - & - & & - & - & - \\
0 & 0 & 0 & 0 & 1 & 0 & -2 & -4
\end{array}\right]
\begin{array}{l}
\#1 \\ \#2 \\ - \\ \#3 \\ \#4 \\ \#5 \\ -
\end{array}
\, , \qquad (2.107)
$$

$$\underbrace{}_{x\ \ y}\ \underbrace{}_{s_1\ s_2}\ \underbrace{}_{s_3\ s_4\ s_5}$$

$$
\begin{bmatrix}
0 & 0 & | & 1 & 0 & | & 0 & -1 & 1 & | & 1 \\
0 & 1 & | & 0 & 0 & | & 0 & 1 & 0 & | & 2 \\
- & - & | & - & - & | & - & - & - & | & - \\
1 & 0 & | & 0 & 0 & | & 0 & -1 & 1 & | & 1 \\
0 & 0 & | & 0 & 0 & | & 1 & 1 & -1 & | & 1 \\
0 & 0 & | & 0 & 1 & | & 0 & 1 & 0 & | & 2 \\
- & - & | & - & - & | & - & - & - & | & - \\
0 & 0 & | & 0 & 0 & | & 0 & -1 & -1 & | & -5
\end{bmatrix}
\begin{array}{l}
\#1 \\ \#2 \\ \\ \#3 \\ \#4 \\ \#5 \\ \\
\end{array}
\qquad (2.108)
$$

$$
\begin{array}{ccccccc}
x & y & s_1 & s_2 & s_3 & s_4 & s_5
\end{array}
$$

We have rederived the conclusion that the corner at $x = 1, y = 2$ maximizes the objective at the value (+)5.

2.9   Further Refining the Simplex Algorithm

In this section we are going to describe an additional, minor, improvement to the Simplex implementation described in Sect. 2.8. (The more profound issue of finding the first, legitimate, corner will be dealt with in the next section.)[26]

We have applied the streamlined Simplex procedure to both Examples 2.1 and 2.2, and we have noted that the *first* step is identical to that of the tried-and-true version (Sect. 2.6)—to wit, create a row-reduced tableau by pivoting in the constraint tableau to install identity columns above the coordinate variables x and y (and z in Example 2.1). But the reader may have noticed that these leading columns never change thereafter; they remain uncluttered. The reason is that the coordinate variables, rendered inactive by this initial pivoting, never become activated again; all of the c-row and s/t stratagems only activate *slack* variables. *Therefore we can—and shall—delete the coordinate-variable columns from the subsequent row-reduced tableaux.*

Moreover, we can also delete the *rows* corresponding to the constraints that are active at the initial corner (i.e. the rows anchored by the **1**'s in the pivoted coordinate-variable columns), because a glance at the row-reduced templates in ((2.93), Sect. 2.6 and (2.75), Sect. 2.7) reveals that the s/t competitions ignore these rows. True, these rows become altered by the subsequent pivoting, but their data are never consulted for the later pivoting options. *We can—and do—delete the constraint rows anchored by the 1's in the coordinate-variable columns, as well.*

Well, a slight correction: these rows display the coordinate values for the corner under consideration, in their final column ((2.93), Sect. 2.6 and (2.75), Sect. 2.7). If we delete them and proceed with a truncated tableau containing only the slack variables, we'll be able to read off the maximum objective function's value ($c \cdot R$), but we won't have the

---

[26] (At long last).

optimum corner coordinates. However that's easy to fix; we'll reveal the secret at the end of Example 2.3.

Example 2.3 Rework Example 2.2 using the truncated Simplex tableaux, beginning at the corner $x = 2$, $y = 1$).

Fig. 2.14 (repeated)
Maximize $x + 2y$ in the shaded pentagon

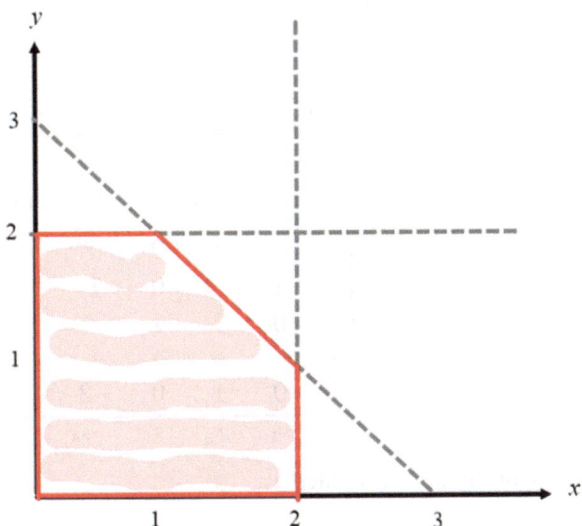

Solution. The constraints and initial tableau are repeated below:

$$\begin{cases} \text{\#1 } -x + s_1 = 0 & active \, (s_1 = 0) \\ \text{\#2 } -y + s_2 = 0 & active \, (s_2 = 0) \\ \text{\#3 } x + s_3 = 2 & inactive \, (s_3 = 2) \,, \quad (2.76, \text{ Sect. } 2.7) \\ \text{\#4 } x + s_4 = 2 & inactive \, (s_4 = 2) \\ \text{\#5 } x + y + s_5 = 3 & inactive \, (s_5 = 3). \end{cases}$$

$$\left.\begin{bmatrix} -1 & 0 & | & 1 & 0 & | & 0 & 0 & 0 & | & 0 \\ 0 & -1 & | & 0 & 1 & | & 0 & 0 & 0 & | & 0 \\ - & - & | & - & - & | & - & - & - & | & - \\ 1 & 0 & | & 0 & 0 & | & 1 & 0 & 0 & | & 2 \\ 0 & 1 & | & 0 & 0 & | & 0 & 1 & 0 & | & 2 \\ 1 & 1 & | & 0 & 0 & | & 0 & 0 & 1 & | & 3 \\ - & - & | & - & - & | & - & - & - & | & - \\ 1 & 2 & | & 0 & 0 & | & 0 & 0 & 0 & | & 0 \end{bmatrix}\begin{matrix} \text{\#1} \\ \text{\#2} \\ - \\ \text{\#3} \\ \text{\#4} \\ \text{\#5} \\ - \\ \end{matrix}\right\} . \, (2.77, \text{ Sect. } 2.7)$$

$$ x \quad y \qquad s_1 \quad s_2 \qquad s_3 \quad s_4 \quad s_5$$

The starting corner lies at the intersection of the third and fifth constraint. To install the proper identity columns, we pivot on the two highlighted 1's:

$$
\left[\begin{array}{cc|cc|ccc|c}
0 & 0 & 1 & 0 & 1 & 0 & 0 & 2 \\
0 & 0 & 0 & 1 & -1 & 0 & 1 & 1 \\
\hline
1 & 0 & 0 & 0 & 1 & 0 & 0 & 2 \\
0 & 0 & 0 & 0 & 1 & 1 & -1 & 1 \\
0 & 1 & 0 & 0 & -1 & 0 & 1 & 1 \\
\hline
0 & 0 & 0 & 0 & 1 & 0 & -2 & -4
\end{array}\right]
\begin{array}{l}
\#1 \\ \#2 \\ - \\ \#3 \\ \#4 \\ \#5 \\ - \\ \;
\end{array}
\tag{2.109}
$$

$$
\underbrace{\;}_{x}\;\underbrace{\;}_{y}\quad \underbrace{\;}_{s_1}\;\underbrace{\;}_{s_2}\quad \underbrace{\;}_{s_3}\;\underbrace{\;}_{s_4}\;\underbrace{\;}_{s_5}
$$

We delete the first two columns and the third and fifth rows.

$$
\left[\begin{array}{ccccc|c}
1 & 0 & 1 & 0 & 0 & 2 \\
0 & 1 & -1 & 0 & 1 & 1 \\
0 & 0 & 1 & 1 & -1 & 1 \\
\hline
0 & 0 & 1 & 0 & -2 & -4
\end{array}\right]
\begin{array}{l}
\#1 \\ \#2 \\ \#4 \\ \;
\end{array}
\tag{2.110}
$$

$$
\underbrace{\;}_{s_1}\;\underbrace{\;}_{s_2}\;\underbrace{\;}_{s_3}\;\underbrace{\;}_{s_4}\;\underbrace{\;}_{s_5}
$$

Pivot on the highlighted 1:

$$
\left[\begin{array}{ccccc|c}
1 & 0 & 0 & -1 & 1 & 1 \\
0 & 1 & 0 & 1 & 0 & 2 \\
0 & 0 & 1 & 1 & -1 & 1 \\
\hline
0 & 0 & 0 & -1 & -1 & -5
\end{array}\right]
\begin{array}{l}
\#1 \\ \#2 \\ \#4 \\ \;
\end{array}
\tag{2.111}
$$

$$
\underbrace{\;}_{s_1}\;\underbrace{\;}_{s_2}\;\underbrace{\;}_{s_3}\;\underbrace{\;}_{s_4}\;\underbrace{\;}_{s_5}
$$

The calculation is finished, and the maximum $\boldsymbol{c}\cdot\boldsymbol{R}$ is (+)5, in agreement with Sect. 2.7. According to the *basic solution* interpretation, the slack variables are $s_1 = 1$, $s_2 = 2$, $s_3 = 1$, $s_4 = s_5 = 0$.

How do we calculate the coordinates of the corner at the maximum objective? We require a formula for x and y in terms of the slack variables. But look at tableau (2.109); the rows we omitted (#3, 5) are just what we need. They state

$$
\begin{cases}
1x + 0y + 0s_1 + 0s_2 + 1s_3 + 0s_4 + 0s_5 = 2 \\
0x + 1y + 0s_1 + 0s_2 - 1s_3 + 0s_4 + 1s_5 = 1
\end{cases},
\tag{2.112}
$$

or $x = 1$, $y = 2$. Again, in agreement with Sect. 2.7.

In this example the full tableau (2.109) contained 48 entries while the truncated ones (2, 3) contained only 24. Such economy may be advantageous in large scale problems requiring many iterations.

In situations where the coordinate variables are constrained to be nonnegative ($x \geq 0$, $y \geq 0$) and the starting corner is the origin, the active constraint equations include

$$-x + s_1 = 0, -y + s_2 = 0, \quad \text{or} \quad x = s_1, y = s_2. \tag{2.113}$$

Since (2.113) directly provides such trivial formulas for x and y in terms of the slack variables, we can bypass the final "trick" calculation (2.112). Indeed, by substituting (s_1, s_2) for (x, y) in the remaining constraints and the objective function we get ab initio a truncated tableau in row-reduced form, ready for Simplex pivoting. An example will demonstrate:

Example 2.4 Rework Example 2.2 using a truncated Simplex tableaux, beginning at the corner $x = 0$, $y = 0$).

Fig. 2.14 (repeated)
Maximize $x + 2y$ in the shaded pentagon

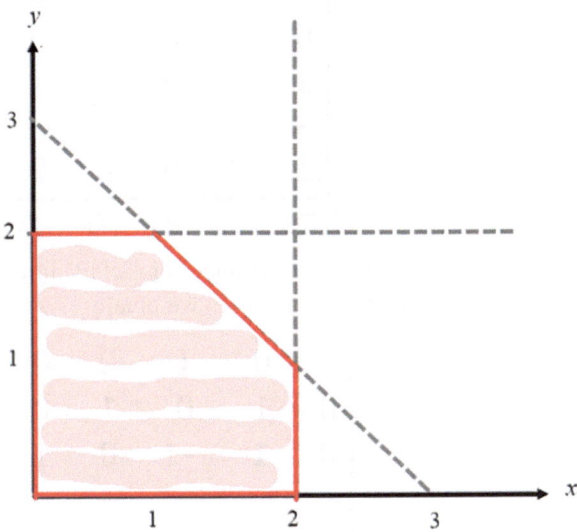

Solution. The constraints

$$
\begin{cases}
\#1 & -x + s_1 = 0 \quad active\ (s_1 = 0) \\
\#2 & -y + s_2 = 0 \quad active\ (s_2 = 0) \\
\#3 & x + s_3 = 2 \quad inactive\ (s_3 = 2) \quad (2.76,\ \text{Sect. } 2.7) \\
\#4 & x + s_4 = 2 \quad inactive\ (s_4 = 2) \\
\#5 & x + y + s_5 = 3\ inactive\ (s_5 = 3).
\end{cases}
$$

enable the substitution (2.113), resulting in the coordinate-free restatement

$$
\text{maximize } s_1 + 2s_2 \text{ subject to}
$$
$$
s_1 + s_3 = 2
$$
$$
s_2 + s_4 = 2
$$
$$
s_1 + s_2 + s_5 = 3.
$$

The truncated tableau for this problem is

$$
\begin{bmatrix}
\mathbf{1} & 0 & 1 & 0 & 0 & | & 2 \\
0 & 1 & 0 & 1 & 0 & | & 2 \\
1 & 1 & 0 & 0 & 1 & | & 3 \\
- & - & - & - & - & & - \\
1 & 2 & 0 & 0 & 0 & | & 0
\end{bmatrix},
\qquad (2.114)
$$
$$
\underbrace{}
$$
$$
s_1{=}x \quad s_2{=}y \quad s_3 \quad s_4 \quad s_5
$$

where we have slipped in a reminder of (2.113) in the caption). Proceeding with Simplex pivoting on the highlighted entries we obtain

$$
\begin{bmatrix}
1 & 0 & 1 & 0 & 0 & | & 2 \\
0 & 1 & 0 & 1 & 0 & | & 2 \\
0 & \mathbf{1} & -1 & 0 & 1 & | & 1 \\
- & - & - & - & - & & - \\
0 & 2 & -1 & 0 & 0 & | & -2
\end{bmatrix},
$$
$$
s_1{=}x \quad s_2{=}y \quad s_3 \quad s_4 \quad s_5
$$

$$
\begin{bmatrix}
1 & 0 & 1 & 0 & 0 & | & 2 \\
0 & 0 & 1 & 1 & -1 & | & 1 \\
0 & 1 & -1 & 0 & 1 & | & 1 \\
- & - & - & - & - & - & - \\
0 & 0 & 1 & 0 & -2 & | & -4
\end{bmatrix} \text{, and}
$$

$$\underbrace{}_{s_1=x \ \ s_2=y \quad s_3 \quad s_4 \qquad s_5}$$

$$
\begin{bmatrix}
1 & 0 & 0 & -1 & 1 & | & 1 \\
0 & 0 & 1 & 1 & -1 & | & 1 \\
0 & 1 & 0 & 1 & 0 & | & 2 \\
- & - & - & - & - & - & - \\
0 & 0 & 0 & -1 & -1 & | & -5
\end{bmatrix} . \qquad (2.115)
$$

$$\underbrace{}_{s_1=x \ \ s_2=y \quad s_3 \quad s_4 \qquad s_5}$$

The basic solution interpretation dictates $x(=s_1) = 1$, $y(=s_2) = 2$ is the maximal corner and $(+)5$ is the objective—in agreement with Sect. 2.7.

If a problem contains coordinate variables that are not restricted to be nonnegative, it may be desirable to reformulate it so as to enable this mechanism. For example Sect. 2.2 pointed out that *any* variable can be expressed as a difference between *two* nonnegative variables. We illustrate this stratagem with Examples 2.5 and 2.6.

Example 2.5 Maximize the objective function $y - 2x$ in the convex polygon (Fig. 2.15) defined by the constraints

$$ y \geq 0, \ x \leq 2, \ y \leq 2, \ x + y \leq 3. \qquad (2.116) $$

Solution. There is no nonnegativity[27] constraint on x, so we introduce $x = x_1 - x_2$ as above, and the reformulation becomes a *three*-dimensional problem:

Maximize the objective function $y - 2(x_1 - x_2)$ subject to the constraints

$$
\begin{array}{llll}
\text{\#1} & x_1 \geq 0 & (-x_1 + s_1 = 0), & \\
\text{\#2} & x_2 \geq 0, & (-x_2 + s_2 = 0), & \\
\text{\#3} & y \geq 0 & (-y + s_3 = 0), & \\
\text{\#4} & x_1 - x_2 \leq 2 & (x_1 - x_2 + s_4 = 2), & (2.117) \\
\text{\#5} & y \leq 2 & (y + s_5 = 2), & \\
\text{\#6} & x_1 - x_2 + y \leq 3 & (x_1 - x_2 + y + s_6 = 3). &
\end{array}
$$

---

[27] A *triple negative*! Miss Slater would be blown away!

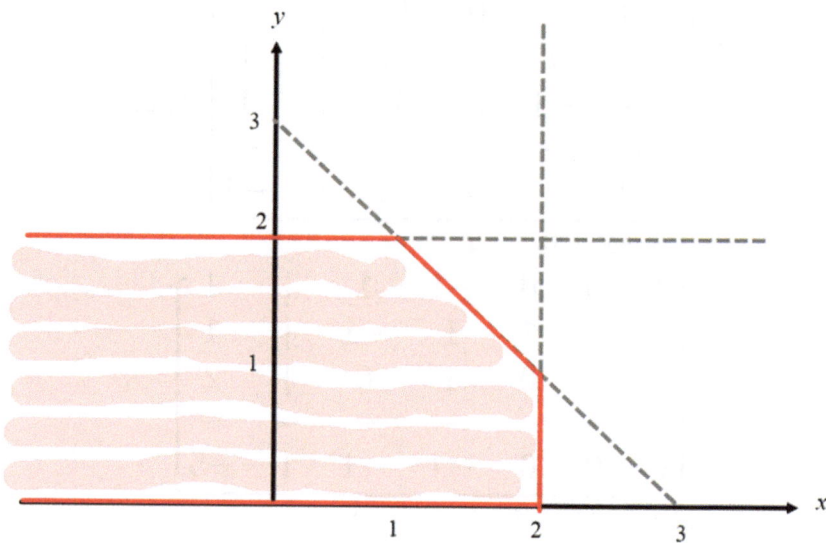

Fig. 2.15 Example 2.4 polygon

Absorbing the first three constraints into the others and into the objective function ($y - 2[x_1 - x_2] = s_3 - 2s_1 + 2s_2$) we obtain the initial tableau

$$
\begin{bmatrix}
1 & -1 & 0 & 1 & 0 & 0 & | & 2 \\
0 & 0 & 1 & 0 & 1 & 0 & | & 2 \\
1 & -1 & 1 & 0 & 0 & 1 & | & 3 \\
- & - & - & - & - & - & - & - \\
-2 & 2 & 1 & 0 & 0 & 0 & | & 0
\end{bmatrix}
\begin{matrix}
\text{\#4} \\
\text{\#5} \\
\text{\#6}
\end{matrix}
\qquad (2.118)
$$

$$\underbrace{\quad}_{s_1=x_1}\ \underbrace{\quad}_{s_2=x_2}\ \underbrace{\quad}_{s_3=y}\ \underbrace{\quad}_{s_4}\ \underbrace{\quad}_{s_5}\ \underbrace{\quad}_{s_6}$$

The basic solution is $s_1 = s_2 = s_3 = 0$, $s_4 = s_5 = 2, s_6 = 3$, legitimately characterizing a corner in x_1, x_2, y-space. (Although $x = 0, y = 0$ is *not* a corner in the two-dimensional Fig. 2.15!).

The **c**-row reveals that we can increase the objective function by deactivating s_2 or s_3; let's try s_3. The s/t winner in the third column is "**1**", the (2, 3) entry. Pivoting thereupon we update the tableau to find

$$
\begin{bmatrix}
1 & -1 & 0 & 1 & 0 & 0 & | & 2 \\
0 & 0 & 1 & 0 & 1 & 0 & | & 2 \\
1 & -1 & 0 & 0 & -1 & 1 & | & 1 \\
- & - & - & - & - & - & - & - \\
-2 & 2 & 0 & 0 & -1 & 0 & | & -2
\end{bmatrix}
\begin{array}{l} \text{\#4} \\ \text{\#5} \\ \text{\#6} \\ \\ \end{array}
\qquad (2.119)
$$

$$s_1 = x_1 \quad s_2 = x_2 \quad s_3 = y \quad s_4 \quad s_5 \quad s_6$$

placing us at $x = s_1 - s_2 = 0 - 0 = 0$, $y = s_3 = 2$, and indicating that increasing s_2 will further increase the objective function.

Now none of the entries in the first three rows of the s_2 column are positive, so the polyhedron is unbounded; there is no limit to how much we can increase $s_2 \equiv x_2$, or decrease $x = (0) - s_2$. Therefore the objective $y - 2x$ has no maximum. (And this is clear from Fig. 2.15).

If we had chosen to try increasing s_2 instead of s_3, we confront the nonpositive entries in column 2 immediately—confirming what we just said before; the polyhedron is unbounded in this direction.

Example 2.6 highlights a minor wrinkle that occurs due to the ambiguity in the values of x_1 and x_2.

Example 2.6 Maximize the objective function $y - 2x$ in the convex polyhedron (Fig. 2.16) defined by the constraints

$$y \geq 0, \ x \leq 2, \quad y \leq 2, \quad x + y \leq 3, \quad x \geq -1. \qquad (2.120)$$

Solution. Since $x \geq 0$ does not appear in the constraints we introduce $x = x_1 - x_2$ as above, and the reformulation becomes:

Maximize the objective function $y - 2(x_1 - x_2)$ subject to the constraints

$$
\begin{array}{lll}
\text{\#1} & x_1 \geq 0 & (-x_1 + s_1 = 0), \\
\text{\#2} & x_2 \geq 0, & (-x_2 + s_2 = 0), \\
\text{\#3} & y \geq 0 & (-y + s_3 = 0) \\
\text{\#4} & x_1 - x_2 \leq 2 & (x_1 - x_2 + s_4 = 2), \\
\text{\#5} & y \leq 2 & (y + s_5 = 2), \\
\text{\#6} & x_1 - x_2 + y \leq 3 & (x_1 - x_2 + y + s_6 = 3), \\
\text{\#7} & x_1 - x_2 \geq -1 & (-x_1 + x_2 + s_7 = 1).
\end{array}
\qquad (2.121)
$$

Adding another constraint (with its slack variable) to the previous initial tableau (2.118), we have

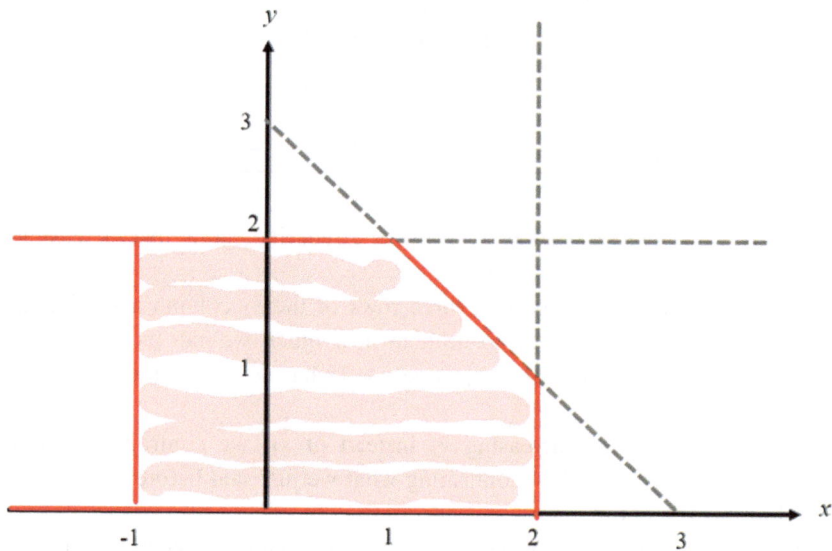

Fig. 2.16 Example 2.5 polygon

$$
\left[
\begin{array}{ccccccc|c}
1 & -1 & 0 & 1 & 0 & 0 & 0 & 2 \\
0 & 0 & 1 & 0 & 1 & 0 & 0 & 2 \\
1 & -1 & 1 & 0 & 0 & 1 & 0 & 3 \\
-1 & 1 & 0 & 0 & 0 & 0 & 1 & 1 \\
\hline
-2 & 2 & 1 & 0 & 0 & 0 & 0 & 0
\end{array}
\right]
\begin{array}{l}
\#4 \\
\#5 \\
\#6 \\
\#7
\end{array}
\qquad (2.122)
$$

$$\underbrace{\qquad}_{s_1=x_1\ s_2=x_2\ s_3=y\quad s_4\quad s_5\quad s_6\quad s_7}$$

Pivoting as indicated, we compute the updates to be

$$
\left[
\begin{array}{ccccccc|c}
0 & 0 & 0 & 1 & 0 & 0 & 1 & 3 \\
0 & 0 & 1 & 0 & 1 & 0 & 0 & 2 \\
0 & 0 & 1 & 0 & 0 & 1 & 1 & 4 \\
-1 & 1 & 0 & 0 & 0 & 0 & 1 & 1 \\
\hline
0 & 0 & 1 & 0 & 0 & 0 & -2 & -2
\end{array}
\right]
\begin{array}{l}
\#4 \\
\#5 \\
\#6 \\
\#7
\end{array}
\qquad \text{and}
$$

$$\underbrace{\qquad}_{s_1=x_1\ s_2=x_2\ s_3=y\quad s_4\quad s_5\quad s_6\quad s_7}$$

$$
\left[
\begin{array}{ccccccc|c}
0 & 0 & 0 & 1 & 0 & 0 & 1 & 3 \\
0 & 0 & 1 & 0 & 1 & 0 & 0 & 2 \\
0 & 0 & 0 & 0 & -1 & 1 & 1 & 2 \\
-1 & 1 & 0 & 0 & 0 & 0 & 1 & 1 \\
\hline
0 & 0 & 0 & 0 & -1 & 0 & -2 & -4
\end{array}
\right]
\begin{array}{l}
\left.\begin{array}{l} \#4 \\ \#5 \\ \#6 \\ \#7 \end{array}\right\} \\ \\ \\
\end{array}
\qquad (2.123)
$$

$$
s_1 = x_1 \; s_2 = x_2 \; s_3 = y \; s_4 \quad s_5 \quad s_6 \quad s_7
$$

The **c**-row in (2.123) indicates that the objective function has reached its maximum, $(+)4$. The basic solution interpretation observes the identity columns and assigns the values $s_2 = 1, s_3 = 2, s_4 = 3, s_6 = 2$. The cluttered columns (which include s_1, according to Sect. 2.3) demand $s_1 = s_5 = s_7 = 0$. Thus the maximum objective in the artificial x_1, x_2, y space occurs at the corner $x_1 = s_1 = 0, x_2 = s_2 = 1, y = s_3 = 2$; in x, y-space the optimal corner lies at $x = x_1 - x_2 = -1, y = s_3 = 2$.

Although this answer solves the problem, it is of interest to note that a situation mentioned in the footnote below Eq. (2.33, Sect. 2.4) has occurred. It is true that no entries in the **c**-row corresponding to the active slack variables $s_1, s_5,$ or s_7 are positive; we can't increase the objective function $\boldsymbol{c} \cdot \boldsymbol{R}$ by deactivating any of these. However the entry in the s_1 column is *zero*, so we can increase $s_1 (\equiv x_1)$ without *lowering* $\boldsymbol{c} \cdot \boldsymbol{R}$. In fact the only constraint equation in (2.123) that involves x_1 is #7, and it remains intact if an increase in x_1 is balanced by an equal decrease in x_2. The polyhedron is unbounded in x_1, x_2, y space, and the objective maximum is achieved everywhere along the edge, extending from the corner $x_1 = 0, x_2 = 1, y = 2$, that maintains $x_1 - x_2 = -1$.

Finally, Sect. 2.2 mentioned that an *equality* constraint like $cx + dy = e$ may be rewritten in the format $cx + dy \le e$ *and* $-cx - dy \le -e$ (to be compatible with Sect. 2.4). But the Simplex formulation can accommodate equality constraints directly—we simply leave out its slack variable. Example 2.7 demonstrates.

Example 2.7 Maximize the objective function $x + 2y$ in the convex region defined by the constraints

$$
x \ge 0, \quad y \ge 0, \quad x \le 2, \quad y \le 2, \quad x + y = 3.
$$

Solution. Figure 2.17 shows that the equality constraint has reduced the problem to one dimension. It's quite trivial but we'll solve it using the Simplex machinery to demonstrate the logic. We formulate the constraints as

$$
\begin{array}{lll}
\#1 & -x + s_1 = 0 & \\
\#2 & -y + s_2 = 0 & \\
\#3 & x + s_3 = 2 & \qquad (2.124) \\
\#4 & y + s_4 = 2 & \\
\#5 & x + y = 3 & (no\,slack\,variable).
\end{array}
$$

Fig. 2.17 Example 2.7

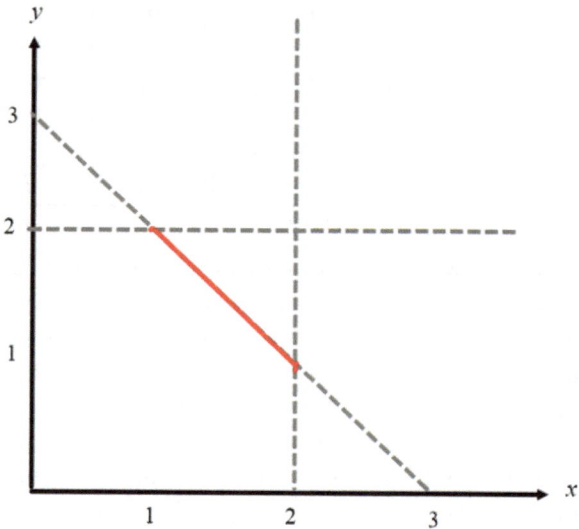

The objective function $c \cdot R = x + 2y$ so c equals $[12]$. The constraint tableau is

$$
\begin{bmatrix}
-1 & 0 & 1 & 0 & 0 & 0 & | & 0 \\
0 & -1 & 0 & 1 & 0 & 0 & | & 0 \\
1 & 0 & 0 & 0 & 1 & 0 & | & 2 \\
0 & 1 & 0 & 0 & 0 & 1 & | & 2 \\
1 & 1 & 0 & 0 & 0 & 0 & | & 3 \\
- & - & - & - & - & - & - & - \\
1 & 2 & 0 & 0 & 0 & 0 & | & 0
\end{bmatrix}
\begin{matrix}
\#1 \\
\#2 \\
- \\
\#3 \\
\#4 \\
\#5 \\
- \\
\end{matrix}
\qquad (2.125)
$$

$$
\underbrace{}_{x} \quad \underbrace{}_{y} \quad \underbrace{}_{s_1} \quad \underbrace{}_{s_2} \quad \underbrace{}_{s_3} \quad \underbrace{}_{s_4}
$$

From Fig. 2.17 it is clear that the intersection $x = y = 0$ is not a legitimate corner. But the intersection of constraint line #3 and constraint #5 *does* lie on the (flat) "polygon", so we row-reduce by pivoting on the highlighted 1's:

$$
\begin{bmatrix}
0 & 0 & 1 & 0 & 1 & 0 & | & 2 \\
0 & 0 & 0 & 1 & -1 & 0 & | & 1 \\
1 & 0 & 0 & 0 & 1 & 0 & | & 2 \\
0 & 0 & 0 & 0 & 1 & 1 & | & 1 \\
0 & 1 & 0 & 0 & -1 & 0 & | & 1 \\
- & - & - & - & - & - & - & - \\
0 & 0 & 0 & 0 & 1 & 0 & | & -4
\end{bmatrix}
\begin{matrix}
\#1 \\
\#2 \\
\#3 \\
\#4 \\
\#5 \\
- \\
\end{matrix}
\qquad (2.126)
$$

$$
\underbrace{}_{x} \quad \underbrace{}_{y} \quad \underbrace{}_{s_1} \quad \underbrace{}_{s_2} \quad \underbrace{}_{s_3} \quad \underbrace{}_{s_4}
$$

We delete the first two columns and the third and fifth rows, and pivot on the highlighted 1 (the s/t winner):

$$
\begin{bmatrix}
1 & 0 & | & 0 & -1 & | & 1 \\
0 & 1 & | & 0 & 1 & | & 2 \\
0 & 0 & | & 1 & 1 & | & 1 \\
- & - & | & - & - & - & - \\
0 & 0 & | & 0 & -1 & | & -5
\end{bmatrix}
\begin{array}{l}
\#1 \\
\#2 \\
\#4 \\
-
\end{array} .
\qquad (2.127)
$$

$$
\underbrace{}_{s_1 \quad s_2} \qquad \underbrace{}_{s_3 \quad s_4}
$$

The maximum objective is (+)5, occurring at the "corner"

$$
x = 2 - 1 s_3 = 1 \quad (\text{deleted row 3 in (5)}),
$$
$$
y = 1 + 1 s_3 = 2 \quad (\text{deleted row 5 in (5)}).
$$

This is hardly a surprise; the only corners were $(1, 2)$ and $(2, 1)$. But it does demonstrate that the Simplex procedure easily accommodates equality constraints without breaking them up into pairs of inequality constraints.

2.10 Phase 1: Finding the First Corner

Since the Simplex algorithm governs a corner-to-corner search for the maximum of the objective function, we must have a procedure for obtaining a legitimate corner of the convex polyhedral search region (the diamond) from which to initiate the search. In our examples so far the polyhedrals were diamond shapes lying in 2- or 3-dimensional space, and the corners were obvious from the figures. But most practical optimization problems involve many more coordinates, and the corners are not evident. We want to stay with two-dimensional examples to keep the arithmetic in check; but for Example 2.8 the reader is admonished not to cheat by skipping ahead and looking at Fig. 2.18.

Remember a (legitimate) corner's properties: it is formed by the intersection of two constraint *equalities*, its active slack variables are zero, and the inactive ones are nonnegative. The constraint submatrix of its truncated row-reduced tableau has a nonnegative final column and its other columns include the identity matrix.

Example 2.8 Maximize the objective function $x + 2y$ in the convex pentagon defined by the constraints

$$
x \geq 0, y \geq 0, x \leq 4, x + y \geq 2, -x + y \leq -1.
$$

Fig. 2.18 Example 2.5

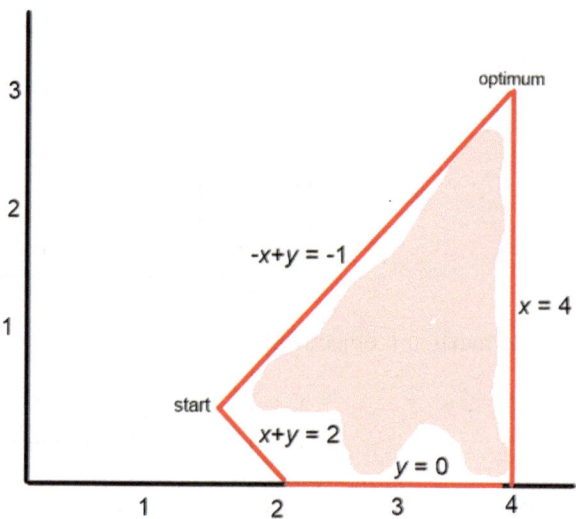

Solution. Introducing slack variables we formulate the constraints as

$$
\begin{aligned}
\#1 \quad & -x + s_1 = 0 \\
\#2 \quad & -y + s_2 = 0 \\
\#3 \quad & x + s_3 = 4 \\
\#4 \quad & -x - y + s_4 = -2 \\
\#5 \quad & -x - y + s_5 = -1,
\end{aligned}
\tag{2.128}
$$

and the constraint tableau is

$$
\left[
\begin{array}{ccccccc|c}
\mathbf{-1} & 0 & 1 & 0 & 0 & 0 & 0 & 0 \\
0 & \mathbf{-1} & 0 & 1 & 0 & 0 & 0 & 0 \\
1 & 0 & 0 & 0 & 1 & 0 & 0 & 4 \\
-1 & -1 & 0 & 0 & 0 & 1 & 0 & -2 \\
-1 & 1 & 0 & 0 & 0 & 0 & 1 & -1 \\
\hline
1 & 2 & 0 & 0 & 0 & 0 & 0 & 0
\end{array}
\right]
\begin{array}{l}
\#1 \\
\#2 \\
\#3 \\
\#4 \\
\#5 \\
\\
0
\end{array}
\tag{2.129}
$$

$$
\begin{array}{ccccccc}
x & y & s_1 & s_2 & s_3 & s_4 & s_5
\end{array}
$$

Let's see if the origin—the intersection of the first two constraint equalities—is a legitimate corner. We install a 2-by-2 identity in the upper left corner by pivoting on the highlighted entries:

$$
\begin{bmatrix}
1 & 0 & -1 & 0 & 0 & 0 & 0 & | & 0 \\
0 & 1 & 0 & -1 & 0 & 0 & 0 & | & 0 \\
0 & 0 & 1 & 0 & 1 & 0 & 0 & | & 4 \\
0 & 0 & -1 & -1 & 0 & 1 & 0 & | & -2 \\
0 & 0 & -1 & 1 & 0 & 0 & 1 & | & -1 \\
- & - & - & - & - & - & - & - & - \\
0 & 0 & 1 & 2 & 0 & 0 & 0 & | & 0
\end{bmatrix}
\begin{matrix}
\#1 \\ \#2 \\ \#3 \\ \#4 \\ \#5 \\ \\ \\
\end{matrix}
\qquad (2.130)
$$
$$
\begin{matrix} x & y & s_1 & s_2 & s_3 & s_4 & s_5 \end{matrix}
$$

Comparison with the row-reduced template (2.93, Sect. 2.6) reveals that this intersection is *not* a legitimate corner because of the negative slack variables $s_4 = -2$, $s_5 = -1$. The Simplex algorithm's guarantee to move us from one corner to another is worthless unless we start it from a corner.

Nonetheless the list (2.124) justifies truncating the tableau as described in Sect. 2.9 (replacing x, y by s_1, s_2 and dropping the first two columns and rows):

$$
\begin{bmatrix}
1 & 0 & 1 & 0 & 0 & | & 4 \\
-1 & -1 & 0 & 1 & 0 & | & -2 \\
-1 & 1 & 0 & 0 & 1 & | & -1 \\
- & - & - & - & - & - & - \\
1 & 2 & 0 & 0 & 0 & | & 0
\end{bmatrix}
\begin{matrix}
\#3 \\ \#4 \\ \#5 \\ \\
\end{matrix}
\qquad (2.131)
$$
$$
\begin{matrix} s_1=x & s_2=y & s_3 & s_4 & s_5 \end{matrix}
$$

How does one find a legitimate corner? The *c*-row is irrelevant for this task, so we ignore it. A promising first move would be to multiply the second and third rows by (-1), rendering the final column nonnegative,

$$
\begin{bmatrix}
1 & 0 & 1 & 0 & 0 & | & 4 \\
1 & 1 & 0 & -1 & 0 & | & 2 \\
1 & -1 & 0 & 0 & -1 & | & 1 \\
- & - & - & - & - & - & - \\
\end{bmatrix}
\begin{matrix}
\#3 \\ \#4 \\ \#5 \\ \\
\end{matrix}
\qquad (2.132)
$$
$$
\begin{matrix} s_1=x & s_2=y & s_3 & s_4 & s_5 \end{matrix}
$$

but this has distorted two of the identity columns.

The motive of our next move is to wedge copies of the missing identity columns back into the tableau like this:

$$
\begin{bmatrix}
0 & 0 & 1 & 0 & 1 & 0 & 0 & | & 4 \\
1 & 0 & 1 & 1 & 0 & -1 & 0 & | & 2 \\
0 & 1 & 1 & -1 & 0 & 0 & -1 & | & 1 \\
- & - & - & - & - & - & - & - & - \\
\end{bmatrix}
\begin{matrix}
\#3 \\ \#4 \\ \#5 \\ \\
\end{matrix}
\qquad (2.133)
$$
$$
\begin{matrix} s_{-1}=\alpha & s_0=\beta & s_1=x & s_2=y & s_3 & s_4 & s_5 \end{matrix}
$$

without disrupting the constraints (2.128). We can accomplish this by defining a new, higher-dimensional linear program:

Maximize the objective function $-\alpha - \beta$ subject to $\alpha \geq 0$, $\beta \geq 0$, and the constraints (2.128).

(Clearly the maximum (zero) will occur at $\alpha = \beta = 0$.) The constraint equality list for the new problem is

$$
\begin{aligned}
-\alpha + s_{-1} &= 0 \\
-\beta + s_0 &= 0 \\
-x + s_1 &= 0 \\
-y + s_2 &= 0 \\
x + s_3 &= 4 \\
-x - y + s_4 &= -2 \\
-x + y + s_5 &= -1,
\end{aligned}
\tag{2.134}
$$

prompting the replacement of α, β, x, y by s_{-1}, s_0, s_1, s_2 in the last three constraints. Appending the appropriate $\boldsymbol{c}$-row for this problem we have its (truncated) constraint tableau

$$
\left[
\begin{array}{ccccccc|c}
0 & 0 & 1 & 0 & 1 & 0 & 0 & 4 \\
\mathbf{1} & 0 & 1 & 1 & 0 & -1 & 0 & 2 \\
0 & \mathbf{1} & 1 & -1 & 0 & 0 & -1 & 1 \\
- & - & - & - & - & - & - & - \\
-1 & -1 & 0 & 0 & 0 & 0 & 0 & 0
\end{array}
\right]
\begin{array}{l}
\#3 \\ \#4 \\ \#5 \\ \\ -
\end{array}
\;,
\tag{2.135}
$$

$$
\underbrace{}_{s_{-1}=\alpha} \; \underbrace{}_{s_0=\beta} \; \underbrace{}_{s_1=x} \; \underbrace{}_{s_2=y} \; \underbrace{}_{s_3} \; \underbrace{}_{s_4} \; \underbrace{}_{s_5}
$$

whose constraint submatrix has the desired form (2.133).

Now we're going to solve the new problem using the Simplex algorithm. To start we pivot on the highlighted entries to achieve a row-reduced form:

$$
\left[
\begin{array}{ccccccc|c}
0 & 0 & 1 & 0 & 1 & 0 & 0 & 4 \\
1 & 0 & 1 & 1 & 0 & -1 & 0 & 2 \\
0 & 1 & \mathbf{1} & -1 & 0 & 0 & -1 & 1 \\
- & - & - & - & - & - & - & - \\
0 & 0 & 2 & 0 & 0 & -1 & -1 & 3
\end{array}
\right]
\begin{array}{l}
\#3 \\ \#4 \\ \#5 \\ \\ -
\end{array}
\;.
\tag{2.136}
$$

$$
\underbrace{}_{s_{-1}=\alpha} \; \underbrace{}_{s_0=\beta} \; \underbrace{}_{s_1=x} \; \underbrace{}_{s_2=y} \; \underbrace{}_{s_3} \; \underbrace{}_{s_4} \; \underbrace{}_{s_5}
$$

Next we pivot on the highlighted s/t winners in turn:

$$
\begin{bmatrix}
0 & -1 & 0 & 1 & 1 & 0 & 1 & | & 3 \\
1 & -1 & 0 & 2 & 0 & -1 & 1 & | & 1 \\
0 & 1 & 1 & -1 & 0 & 0 & -1 & | & 1 \\
- & - & - & - & - & - & - & | & - \\
0 & -2 & 0 & 2 & 0 & -1 & 1 & | & 1
\end{bmatrix}
\begin{matrix}
\#3 \\ \#4 \\ \#5 \\ \\ \\
\end{matrix} ,
$$
$$
\begin{matrix}
s_{-1}{=}\alpha & s_0{=}\beta & s_1{=}x & s_2{=}y & s_3 & s_4 & s_5
\end{matrix}
$$

$$
\begin{bmatrix}
-.5 & -.5 & 0 & 0 & 1 & .5 & .5 & | & 2.5 \\
.5 & -.5 & 0 & 1 & 0 & -.5 & .5 & | & .5 \\
.5 & .5 & 1 & 0 & 0 & -.5 & -.5 & | & 1.5 \\
- & - & - & - & - & - & - & | & - \\
-1 & -1 & 0 & 0 & 0 & 0 & 0 & | & 0
\end{bmatrix}
\begin{matrix}
\#3 \\ \#4 \\ \#5 \\ \\ \\
\end{matrix} . \qquad (2.137)
$$
$$
\begin{matrix}
s_{-1}{=}\alpha & s_0{=}\beta & s_1{=}x & s_2{=}y & s_3 & s_4 & s_5
\end{matrix}
$$

The final tableau confirms that the maximum of $-\alpha - \beta$ is zero; hence $\alpha = \beta = 0$ and the columns we inserted into the constraint submatrix can be dropped. The remaining constraints—the original ones in (2.128)—have an equivalent expression

$$
\begin{bmatrix}
0 & 0 & 1 & .5 & .5 & | & 2.5 \\
0 & 1 & 0 & -.5 & .5 & | & .5 \\
1 & 0 & 0 & -.5 & -.5 & | & 1.5 \\
- & - & - & - & - & | & -
\end{bmatrix}
\begin{matrix}
\#3 \\ \#4 \\ \#5 \\ \\
\end{matrix} \qquad (2.138)
$$
$$
\begin{matrix}
s_1{=}x & s_2{=}y & s_3 & s_4 & s_5
\end{matrix}
$$

whose basic solution $x = 1.5, y = .5$ is a legitimate corner for our original problem. Here's a diagram of the polygon:

Now let's reattach our original c-row in (2.131):

$$
\begin{bmatrix}
0 & 0 & 1 & .5 & .5 & | & 2.5 \\
0 & 1 & 0 & -.5 & .5 & | & .5 \\
1 & 0 & 0 & -.5 & -.5 & | & 1.5 \\
- & - & - & - & - & | & - \\
1 & 2 & 0 & 0 & 0 & | & 0
\end{bmatrix}
\begin{matrix}
\#3 \\ \#4 \\ \#5 \\ \\ \\
\end{matrix} . \qquad (2.139)
$$
$$
\begin{matrix}
s_1{=}x & s_2{=}y & s_3 & s_4 & s_5
\end{matrix}
$$

If you pivot on the s/t winners in entries $(3, 1)$, $(2, 2)$, $(1, 4)$, and $(2, 2)$ in turn, you'll arrive at the tableau

$$
\begin{bmatrix}
0 & 0 & 2 & 1 & 1 & | & 5 \\
0 & 1 & 1 & 0 & 1 & | & 3 \\
1 & 0 & 1 & 0 & 0 & | & 4 \\
- & - & - & - & - & | & - \\
0 & 0 & -3 & 0 & -2 & | & -10
\end{bmatrix}
\begin{matrix}
\#3 \\ \#4 \\ \#5 \\ \\ \\
\end{matrix} . \qquad (2.140)
$$
$$
\begin{matrix}
s_1{=}x & s_2{=}y & s_3 & s_4 & s_5
\end{matrix}
$$

The maximum objective for our problem is $(+)10$, and it occurs at the corner $x = 4$, $y = 3$.

The overall process of rendering the final column of the constraint submatrix nonnegative, inserting copies of the damaged identity columns, minimizing the negative sum of its slack variables, and extracting a legitimate starting corner is known as the *Phase 1* procedure of the Simplex Algorithm.

2.11   Degeneracy (Optional Reading)

A corner of a planar convex figure occurs when two supporting lines intersect; in three dimensions, when three supporting planes intersect. When *more* than two lines, or more than three planes, intersect at a corner it is called *degeneracy*. In n-dimensional space degeneracy occurs when more than n supporting hyperplanes pass through a corner.

We have diligently skirted around this issue so far. The fact is, degeneracy is a very low probability occurrence in real-data linear programs. Consider the following:

Question 13. The two lines described by

$$\left.\begin{array}{r} 2.404x + 3.708y = 19.650 \\ 1.524x - 2.428y = -0.087 \end{array}\right\}$$

intersect at the point

$$\begin{cases} x = 4.125 \\ y = 2.625 \end{cases}.$$

If r is a random number, what is the probability that the third line

$$1.008x + 3.258y = r$$

will pass through the same point?

Answer. Clearly r will have to equal

$$r = 1.008 \times 4.125 + 3.258 \times 2.625 = 12.71025.$$

If r is chosen at random from the continuum according to any *continuous* probability density function, the probability that it will take this precise value is zero.

So, for such data, the occurrence of degeneracy is a statistical anomaly. However, when linear programs are contrived for testing[28] the analysts are be inclined to choose data that keep the calculations simple:

---

[28] Or textbook writing.

Question 14. The two lines described by

$$\left.\begin{array}{c} 2x + y = 4 \\ x + 2y = 5 \end{array}\right\}$$

intersect at the point

$$\left\{\begin{array}{l} x = 1 \\ y = 2 \end{array}\right..$$

If r is chosen at random from the set $\{1, 2, 3, 4, 5\}$, what is the probability that the third line

$$x + y = r$$

will pass through the same point?
Answer. r will have to equal 3; the probability of degeneracy is 20%.[29]
But to quote Dantzig,

"It turned out that although the probability of a linear program being degenerate was zero, every practical problem tested by my branch in the Air Force turned out to be so." ("Reminiscences ...", op cit)

So for the sake of credibility, your author will now address the unlikely occurrence of degeneracy.

To get a concrete picture of the degeneracy phenomenon, we return to the diamond analogy. Suppose a diamond cutter starts with a rough diamond (two-dimensional), and cleaves it along the lines sketched in Fig. 2.19.

We seek the highest point in the diamond (obviously $x = 5$, $y = 4$). The linear program formulation is

Maximize y subject to the constraints

$$\left\{\begin{array}{ll} \#1 & x \geq 0 \\ \#2 & y \geq 0 \\ \#3 & x \leq 5 \\ \#4 & y \leq x + 1 \\ \#5 & 2y \leq x + 3 \end{array}\right., \tag{2.141}$$

and we can take the origin $(0, 0)$ as the starting point for the Simplex algorithm. (It is clearly a legitimate corner, and there is no degeneracy.)

---

[29] By the same logic, the probability that a "truly" random square matrix will be singular is zero.

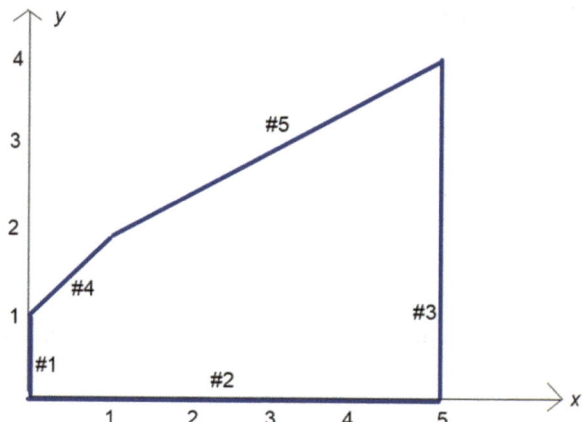

Fig. 2.19 Cleaved diamond

Consult Fig. 2.20 as we conduct a brisk tour of the Simplex actions.

(a) The active edges passing through the original corner correspond to constraints #1 and 2. The objective function y only increases along #1, so we deactivate #2.

(b) We slide along edge #1 until we encounter the next obstacle (constraint), #4, at $x = 0, y = 1$. The edges at this corner correspond to constraints #1 and 4, and the objective will increase only if we proceed along #4—deactivating #1.

(c) The next obstacle along edge #4 is edge #5, which is encountered at $x = 1, y = 2$.

(d) To increase y we deactivate #4 and proceed along #5 until we collide with edge #3 at $x = 5, y = 4$. From this corner neither active edge (#5 or 3) leads to an increase in y, and the algorithm terminates. (No degeneracy, no problem.)

To introduce degeneracy into this scenario, suppose the cutter had decided to cleave the diamond further, along edges #6 and 7 in Fig. 2.21. Lines #5 and 6 "support" the diamond in the sense of Sect. 2.4, but they have been trimmed off. If we retain them among the constraint list, the corner at $(1, 2)$ becomes degenerate because it lies on *four* supporting lines. Of course from the figure we can see that the diamond is actually well-defined by edges #1, 2, 3, 4, and 7. However, in commercial optimization situations where there are thousands of constraints and coordinate variables, it would be impractical to scrutinize each corner for degeneracy and sort out the disenfranchised constraints. So we'll retain all the constraints and formulate this linear program as

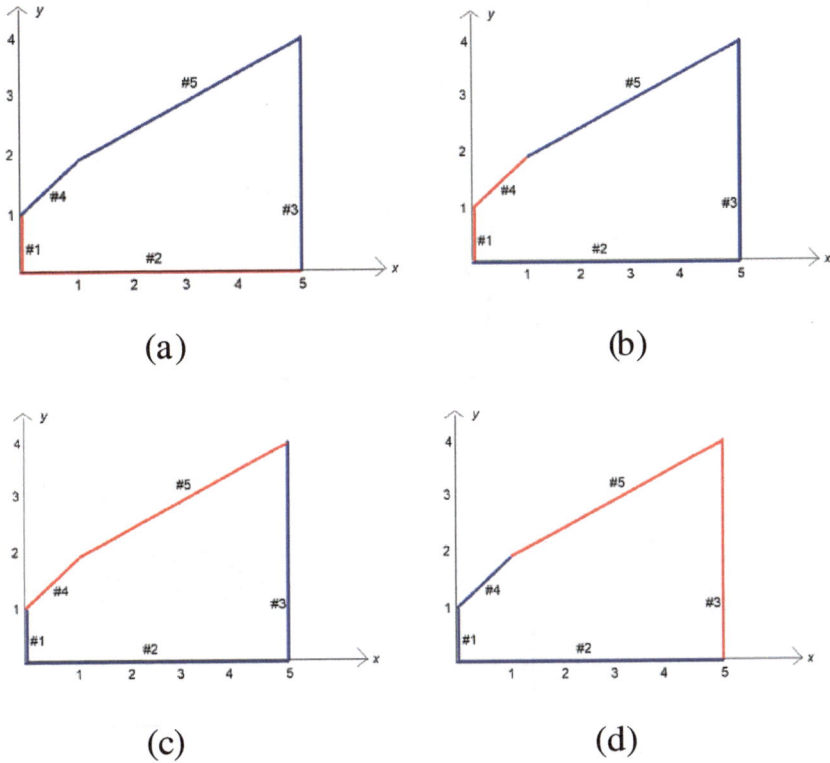

Fig. 2.20 Simplex snapshots

Maximize y subject to the constraints

$$
\begin{cases}
\#1 & x \geq 0 \\
\#2 & y \geq 0 \\
\#3 & x \leq 5 \\
\#4 & y \leq x + 1 \\
\#5 & 2y \leq x + 3 \\
\#6 & 3y \leq x + 5 \\
\#7 & 6y \leq x + 11
\end{cases}
\qquad . \qquad (2.142)
$$

To visualize the Simplex actions for this configuration we'll be a bit whimsical in interpreting Fig. 2.22. Later we'll display the actual tableaux and confirm our fanciful imagery.

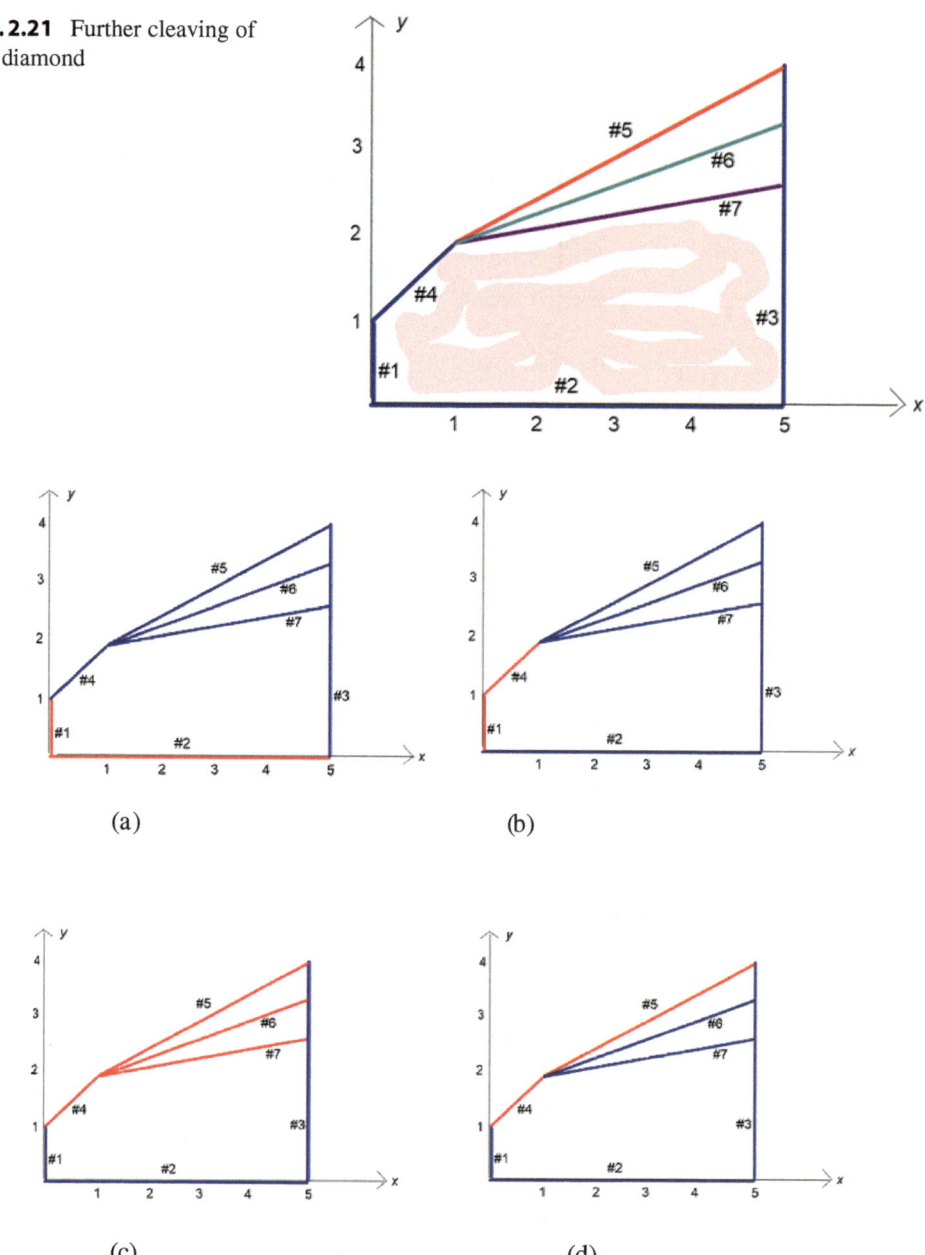

Fig. 2.21 Further cleaving of the diamond

Fig. 2.22 Simplex snapshots: degeneracy

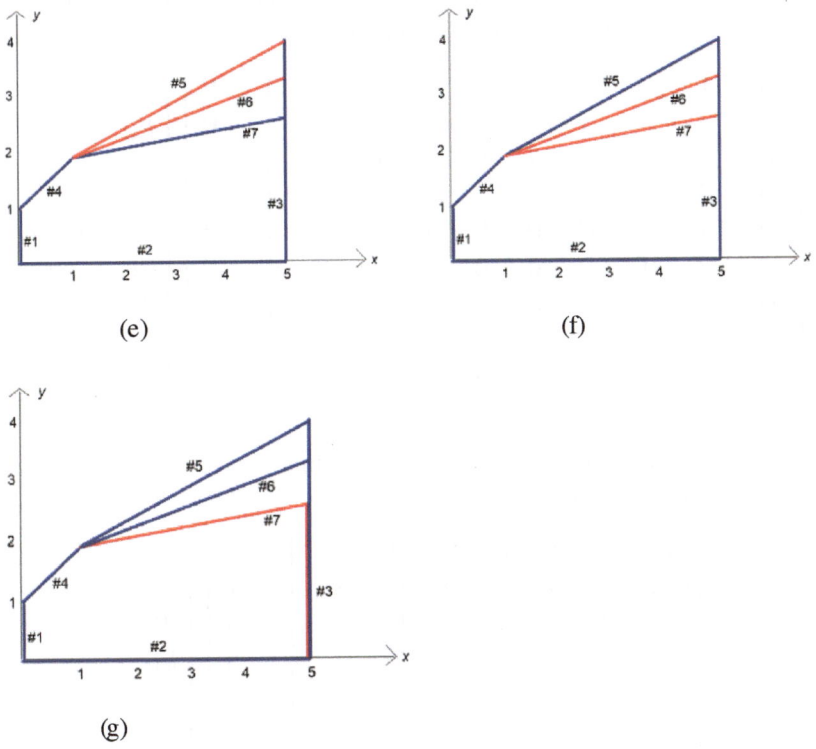

Fig. 2.22 (continued)

(a) The Simplex strategy starts as before, moving from $(0, 0)$ to $(0, 1)$ as it increases the objective function.

(b) At $(0, 1)$ it sees edge #4 as a direction for increasing y, but in probing for the next corner on this edge it foresees a roadblock at $(1, 2)$ simultaneously manned by *three* constraints (#5, 6, and 7).

(c) We need to give Simplex a tiebreaker rule to decide which of the three constraints will be regarded as the barrier stonewalling it at the corner $(1, 2)$.

(d) Suppose, say, that Simplex decides to choose edge #5 as the edge to activate by pivoting. *Note that #6 and 7 will also be activated by the shift to* $(1, 2)$. But Simplex can't advance along edge #5 because constraints #6 and 7 immediately impede it.

(e) Therefore it pivots so as to (redundantly) activate, say, edge #6. Still sitting at $(1, 2)$ and probing edge #6 for the next corner, it immediately encounters constraint #7.

(f) It (again, redundantly) activates edge #7 and probes it for the next corner.

(g) At last it is able to depart $(1, 2)$ and shift to $= 5$, $y = \frac{16}{6} = 2\frac{2}{3}$, which is the maximum point.

Here is the sequence of tableaux for the example. The initial constraint tableau[30] for the objective function y and the constraints (2.142) is

$$
\left[
\begin{array}{rrrrrrrrr|r}
-1 & 0 & 1 & 0 & 0 & 0 & 0 & 0 & 0 & 0 \\
0 & -1 & 0 & 1 & 0 & 0 & 0 & 0 & 0 & 0 \\
1 & 0 & 0 & 0 & 1 & 0 & 0 & 0 & 0 & 5 \\
-1 & 1 & 0 & 0 & 0 & 1 & 0 & 0 & 0 & 1 \\
-1 & 2 & 0 & 0 & 0 & 0 & 1 & 0 & 0 & 3 \\
-1 & 3 & 0 & 0 & 0 & 0 & 0 & 1 & 0 & 5 \\
-1 & 6 & 0 & 0 & 0 & 0 & 0 & 0 & 1 & 11 \\
\hline
0 & 1 & 0 & 0 & 0 & 0 & 0 & 0 & 0 & 0
\end{array}
\right]
\begin{array}{l}
\#1 \\ \#2 \\ \#3 \\ \#4 \\ \#5 \\ \#6 \\ \#7 \\ \\ 0
\end{array}
, \qquad (2.143)
$$

$$
\underbrace{}
$$
$x \quad y \quad s_1 \quad s_2 \quad s_3 \quad s_4 \quad s_5 \quad s_6 \quad s_7$

and with the indicated pivoting and deletion of the first two rows and columns we find the truncated row-reduced version to be

$$
\left[
\begin{array}{rrrrrrr|r}
1 & 0 & 1 & 0 & 0 & 0 & 0 & 5 \\
-1 & 1 & 0 & 1 & 0 & 0 & 0 & 1 \\
-1 & 2 & 0 & 0 & 1 & 0 & 0 & 3 \\
-1 & 3 & 0 & 0 & 0 & 1 & 0 & 5 \\
-1 & 6 & 0 & 0 & 0 & 0 & 1 & 11 \\
\hline
0 & 1 & 0 & 0 & 0 & 0 & 0 & 0
\end{array}
\right]
\begin{array}{l}
\#3 \\ \#4 \\ \#5 \\ \#6 \\ \#7 \\ \\ 0
\end{array}
. \qquad (2.144)
$$

$s_1 \quad s_2 \quad s_3 \quad s_4 \quad s_5 \quad s_6 \quad s_7$

The basic solution for tableau (2.144) expresses the slack variables at the starting corner $(9, 0)$, Fig. 2.22a. The c-row and the s/t competition direct us to pivot on the "1", producing:

$$
\left[
\begin{array}{rrrrrrr|r}
1 & 0 & 1 & 0 & 0 & 0 & 0 & 5 \\
-1 & 1 & 0 & 1 & 0 & 0 & 0 & 1 \\
1 & 0 & 0 & -2 & 1 & 0 & 0 & 1 \\
2 & 0 & 0 & -3 & 0 & 1 & 0 & 2 \\
5 & 0 & 0 & -6 & 0 & 0 & 1 & 5 \\
\hline
1 & 0 & 0 & -1 & 0 & 0 & 0 & -1
\end{array}
\right]
\begin{array}{l}
\#3 \\ \#4 \\ \#5 \\ \#6 \\ \#7 \\ \\ 0
\end{array}
. \qquad (2.145)
$$

$s_1 \quad s_2 \quad s_3 \quad s_4 \quad s_5 \quad s_6 \quad s_7$

s_1 and s_4 are the active variables, placing us at the corner $(0, 1)$ in Fig. 2.22b. The tableau directs us to deactivate s_1, and the minimal eligible s/t value in the first column is 1,

---

[30] Don't dwell on all the data in these tableaux; we'll highlight the relevant entries.

indicating that the next corner on edge #4 is $x = 1$, $y = 2$ (Eq. 2.37, Sect. 2.4). However, there is a tie among the s/t competitors in the first column ($1/1 = 2/2 = 5/5$), so that pivoting on either the "1", "2", or "5" is valid (Fig. 2.22c). Our decision to pivot on the "1" activates constraint #5 and leads to the tableau

$$
\left[
\begin{array}{ccccccc|c}
0 & 0 & 1 & 2 & -1 & 0 & 0 & 4 \\
0 & 1 & 0 & -1 & 1 & 0 & 0 & 2 \\
1 & 0 & 0 & -2 & 1 & 0 & 0 & 1 \\
0 & 0 & 0 & 1 & -2 & 1 & 0 & 0 \\
0 & 0 & 0 & 4 & -5 & 0 & 1 & 0 \\
\hline
0 & 0 & 0 & 1 & -1 & 0 & 0 & -2
\end{array}
\right]
\begin{array}{l}
\#3 \\ \#4 \\ \#5 \\ \#6 \\ \#7 \\ \\ 0
\end{array}
\qquad (2.146)
$$

$$s_1 \quad s_2 \quad s_3 \quad s_4 \quad s_5 \quad s_6 \quad s_7$$

We have arrived at the degenerate corner $(1, 2)$ and increased the objective function to $(+)2$. And for the first time we are seeing **zeros** in the final column of a row-reduced tableau. The basic solution for tableau (2.146) has s_4, s_5, s_6, and s_7 all equal to zero—s_4 and s_5 because their columns are cluttered, and s_6 and s_7 because they anchor identity columns with corresponding zeros in the last column. *The latter zeros are Simplex's way of maintaining five identity columns in the constraint submatrix while signaling that there are four, not two, zero-valued slack variables.*

Tableau (2.146) directs us to deactivate constraint #4 and to replace it with #6 or 7. But once again we are faced with an s/t tie in column 4: $0/1 = 0/4$. Let's pivot on the "1":

$$
\left[
\begin{array}{ccccccc|c}
0 & 0 & 1 & 0 & 3 & -2 & 0 & 4 \\
0 & 1 & 0 & 0 & -1 & 1 & 0 & 2 \\
1 & 0 & 0 & 0 & -3 & 2 & 0 & 1 \\
0 & 0 & 0 & 1 & -2 & 1 & 0 & 0 \\
0 & 0 & 0 & 0 & 3 & -4 & 1 & 0 \\
\hline
0 & 0 & 0 & 0 & 1 & -1 & 0 & -2
\end{array}
\right]
\begin{array}{l}
\#3 \\ \#4 \\ \#5 \\ \#6 \\ \#7 \\ \\ 0
\end{array}
\qquad (2.147)
$$

$$s_1 \quad s_2 \quad s_3 \quad s_4 \quad s_5 \quad s_6 \quad s_7$$

We haven't moved from the corner $(1, 2)$ so the slack variables s_4, s_5, s_6, and s_7 are still zero; now s_5 and s_6 anchor cluttered columns and s_4 and s_7 anchor identity columns. And the objective function is still stalled at $(+)2$. This is depicted in Fig. 2.22e. Pivoting on the highlighted "3" in the tableau we obtain

$$
\begin{bmatrix}
0 & 0 & 1 & 0 & 0 & 2 & -1 & | & 4 \\
0 & 1 & 0 & 0 & 0 & -1/3 & 1/3 & | & 2 \\
1 & 0 & 0 & 0 & 0 & -2 & 1 & | & 1 \\
0 & 0 & 0 & 1 & 0 & -5/3 & 2/3 & | & 0 \\
0 & 0 & 0 & 0 & 1 & -4/3 & 1/3 & | & 0 \\
- & - & - & - & - & - & - & - & - \\
0 & 0 & 0 & 0 & 0 & 1/3 & -1/3 & | & -2
\end{bmatrix}
\begin{matrix}
\#3 \\ \#4 \\ \#5 \\ \#6 \\ \#7 \\ \\ 0
\end{matrix}
\qquad (2.148)
$$

$$
\underset{s_1 \quad s_2 \quad s_3 \quad s_4 \quad s_5 \quad s_6 \qquad s_7}{}
$$

with s_6 and s_7 anchoring cluttered columns and s_4 and s_5 anchoring identity columns; see Fig. 2.22f. Pivoting on the "2" in the tableau will activate constraint #3 and enable us to move off of the corner $(1, 2)$ (and, at last, increase the objective function):

$$
\begin{bmatrix}
0 & 0 & 1/2 & 0 & 0 & 1 & -1/2 & | & 2 \\
0 & 1 & 1/6 & 0 & 0 & 0 & 1/6 & | & 8/3 \\
1 & 0 & 1 & 0 & 0 & 0 & 0 & | & 5 \\
0 & 0 & 5/6 & 1 & 0 & 0 & -1/6 & | & 10/3 \\
0 & 0 & 2/3 & 0 & 1 & 0 & -1/3 & | & 8/3 \\
- & - & - & - & - & - & - & - & - \\
0 & 0 & -1/6 & 0 & 0 & 0 & -1/6 & | & -8/3
\end{bmatrix}
\begin{matrix}
\#3 \\ \#4 \\ \#5 \\ \#6 \\ \#7 \\ \\ 0
\end{matrix}
\qquad (2.149)
$$

$$
\underset{s_1 \quad s_2 \quad s_3 \quad s_4 \quad s_5 \quad s_6 \qquad s_7}{}
$$

We have arrived at the maximum point: $x = 5$, $y = 2\frac{2}{3}$.

Thus we have seen that degeneracy can clog up the Simplex computations. Although in this example it wasn't too bad, the aimless fidgeting forbodes trouble in higher-dimensional situations; and in fact several examples of degenerate problems that idle forever, cycling between corners sharing the same value of the objective function, have turned up. For your amusement we cite one of the simplest (Hall and McKinnon 2000) (rescaled for convenience):

Maximize $46x_1 + 43x_2 - 271x_3 - 8x_4$ subject to the constraints

$$
\begin{cases}
\#1 & 2x_1 + x_2 - 7x_3 - 1x_4 \leq 0 \\
\#2 & -39x_1 - 7x_2 + 39x_3 + 2x_4 \leq 0
\end{cases}
\qquad (2.150)
$$

The truncated row-reduced initial tableau is given by

$$
\begin{bmatrix}
2 & 1 & -7 & -1 & 1 & 0 & | & 0 \\
-39 & -7 & 39 & 2 & 0 & 1 & | & 0 \\
- & - & - & - & - & - & - & - \\
46 & 43 & -271 & -8 & 0 & 0 & | & 0
\end{bmatrix}
\qquad (2.151)
$$

$$
\underset{s_1 \quad s_2 \quad s_3 \quad s_4 \quad s_5 \quad s_6}{}
$$

Fig. 2.23 Realistic version of Fig. 2.22

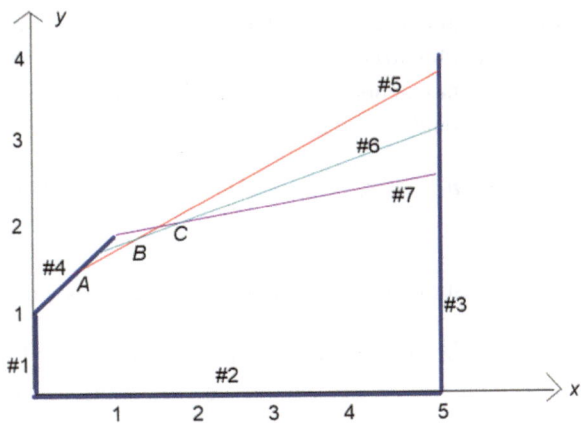

If you break s/t ties by pivoting on the largest t you will find that the tableau reverts to the form (2.151) every six iterations.[31]

That is, as long as you use exact arithmetic. Degeneracy is a very fragile phenomenon. Suppose the diamond cutter failed to execute the last three cleaves in Fig. 2.21 *exactly* through $x - 1, y = 2$. The diamond would then have a shape like Fig. 2.23:

There is no degeneracy, and Simplex would march unimpeded through $(0, 1)$, A, B, and C to the maximum corner $(5, 2\frac{2}{3})$.

In fact on most digital computers the accumulation of roundoff errors in the Simplex iterations would eventually throw the numbers off and shatter the degeneracy (if the user can wait that long).

Some analysts (Charnes 1952; Beale 1955; Bland 1977; Avis et al. 2008) have invented tiebreaker rules that are guaranteed to overcome cycling. Charnes's rule, in fact *exploits* the fragility of the anomaly to deduce the sequence decisions that Simplex would follow if the tableau data were perturbed. It directs us to break s/t ties by choosing, among the tied rows, that row with the lowest ratio of *first*-column entry to t entry. (If there are still ties, we look for the lowest ratio of *second*-column entry to t entry. And so on.) For Hall/McKinnon's example (2.151) the algorithm terminates on the fourth iteration with the tableau

$$
\begin{bmatrix}
5 & 0 & 2 & 1 & -1.4 & -.2 & | & 0 \\
7 & 1 & -5 & 0 & -.4 & -.2 & | & 0 \\
- & - & - & - & - & - & - & - \\
-215 & 0 & -40 & 0 & 6 & 7 & | & 0
\end{bmatrix}.
\qquad (2.152)
$$

$s_1 \quad s_2 \quad s_3 \quad s_4 \quad s_5 \quad s_6$

---

[31] This pivoting strategy is common practice in matrix computations, to avoid numerical instability (Saff, E. B., Snider, A. D. (2016)).

indicating the diamond and the objective function are unbounded. For example the constraint submatrix states that for any positive values of s_5 and s_6, a solution with non-negative components would be $x_1 = x_3 = 0$, $x_2 = .4s_5 + .2s_6$, $x_4 = 1.4s_5 + .2s_6$, and the objective would be the unbounded function

$$\mathbf{c} \cdot \mathbf{R} = 46x_1 + 43x_2 - 271x_3 - 8x_4 = 43(.4s_5 + .2s_6) - 8(1.4s_5 + .2s_6)$$
$$= 6s_5 + 7s_6.$$

The constraints (2.151) are comfortably satisfied:

$$2x_1 + x_2 - 7x_3 - 1x_4 = .4s_5 + .2s_6 - (1.4s_5 + .2s_6)$$
$$= -s_5, \quad -39x_1 - 7x_2 + 39x_3 + 2x_4$$
$$= -7(.4s_5 + .2s_6) + 2(1.4s_5 + .2s_6) = -s_6.$$

2.12   Duality

The concept of duality is very useful in advanced optimization theory. The following example illustrates the point. We revisit the Antarctic expedition example that opened this chapter:

> A dietician predicts that her expedition team should consume 2300 ounces of milk chocolate and 1200 ounces of almonds during an upcoming 10-week Antarctic exposition. Her outfitter can supply her with chocolate almond bars, each containing 1.08 ounces of milk chocolate and 0.36 ounces of almonds, for $1.50 apiece; and he can supply bags of chocolate-covered almonds, each containing 2.75 ounces of chocolate and 2.03 ounces of almonds, for $3.75 each. (Fractional bags and bars are permitted.)

The obvious problem—how many chocolate bars and covered almonds should she buy to meet or exceed the dietary requirements, at minimal cost—will be designated as our "primal" linear program.

> To visualize the *dual* linear program, suppose another outfitter can sell the dietician milk chocolate and almonds *in bulk*. How much should he charge per ounce for the chocolate and almonds, to maximize his profit? Bear in mind that if a chocolate *bar* costs less than the equivalent amount of chocolate and almonds sold separately, the dietician will buy enough *bars* from the first dealer to meet, say, her chocolate needs, and then buy the remaining almonds from the second dealer. Therefore the second dealer's prices must be low enough so that 1.08 ounces of milk chocolate and 0.36 ounces of almonds—the equivalent of a chocolate bar—cost less than $1.50, the price of a bar. Similarly the equivalent of a bag of covered almonds—2.75 ounces of chocolate and 2.03 ounces of almonds—must be cheaper than the cost of the bag, $4.50. The dual linear program, then, seeks the prices that maximize the profit, subject to these cost restrictions.

Let's set up the mathematical formulations of the primal and dual programs.

If the dietician orders x_1 chocolate bars and x_2 covered almonds, she has $1.08x_1 + 2.75x_2$ ounces of chocolate; and she requires at least 2300 oz, so $1.08x_1 + 2.75x_2 \geq 2300$. Similarly the almond requirement is $0.36x_1 + 2.03x_2 \geq 1100$. And her goal is to minimize the total cost $1.50x_1 + 3.75x_2$.

Primal Linear Program

Minimize $1.50x_1 + 3.75x_2$ or $[1.50 \quad 3.75]\begin{bmatrix} x_1 \\ x_2 \end{bmatrix}$.

subject to $1.08x_1 + 2.75x_2 \geq 2300$ and $0.36x_1 + 2.03x_2 \geq 1200$ or

$$\begin{bmatrix} 2300 \\ 1200 \end{bmatrix} \leq \begin{bmatrix} 1.08 & 2.75 \\ 0.36 & 2.03 \end{bmatrix}\begin{bmatrix} x_1 \\ x_2 \end{bmatrix}, \tag{2.153}$$

and

$$x_1 \geq 0, x_2 \geq 0.$$

If the second supplier sets his prices at y_1 dollars per ounce of milk chocolate and y_2 dollars per ounce of almonds, his goal is

Dual Linear Program

maximize $2300y_1 + 1200y_2$ or $[y_1 \quad y_2]\begin{bmatrix} 2300 \\ 1200 \end{bmatrix}$

subject to $1.08y_1 + 0.36y_2 \leq 1.50$ and $2.75y_1 + 2.03y_2 \leq 3.75$ or

$$[y_1 \quad y_2]\begin{bmatrix} 1.08 & 2.75 \\ 0.36 & 2.03 \end{bmatrix} \leq [1.50 \quad 3.75] \tag{2.154}$$

and $y_1 \geq 0, y_2 \geq 0.$

These equations reveal an interesting fact. The coefficient matrices in (2.153) and (2.154) are identical; *and*, since the variables are nonnegative we can left-multiply inequality (2.153) by $[y_1 \quad y_2]$, and right-multiply (2.154) by $\begin{bmatrix} x_1 \\ x_2 \end{bmatrix}$, while maintaining the inequalities. Thus we derive the following inequality chain:

$$[y_1\ y_2]\begin{bmatrix} 2300 \\ 1200 \end{bmatrix} \leq [y_1\ y_2]\begin{bmatrix} 1.08 & 2.75 \\ 0.36 & 2.03 \end{bmatrix}\begin{bmatrix} x_1 \\ x_2 \end{bmatrix} \leq [1.50\ \ 3.75]\begin{bmatrix} x_1 \\ x_2 \end{bmatrix}.$$

Aha! Look at the first and last members; for *any* feasible choice (meeting the nonnegativity constraints) of the variables, the second supplier's profit cannot exceed the cost paid to the first grocer; and therefore the *maximum* profit is less than or equal to the minimum cost. To generalize:

> The objective function for any feasible solution to the dual problem (it does not have to be optimal) provides a lower bound to the objective function for any feasible solution to the primal problem.

This raises the possibility of improving the Simplex algorithm by judiciously switching between the primal and dual programs. The *primal–dual Simplex algorithm* exploits this. We refer the reader to the literature.

References

Avis, D., Kaluzny, B., Titley-Péloquin, D.: Visualizing and constructing cycles in the simplex method. Oper. Res. **56**(2), 512–518 (2008)

Beale, E.: Cycling in the dual simplex method. Naval Res. Logist. Q. **2**(4), 269–275 (1955)

Bland, R.G.: New finite pivoting rules for the simplex method. Math. Oper. Res. **2**, 103–107 (1977)

Charnes, A.: Optimality and degeneracy in linear programming. Econometrica **20**(2), 160–170 (1952)

Charnes, A., Cooper, W.W., Mellon, B.: Blending aviation gases—a study in programming independent activites in an integrated oil company. Econometrica **20**(2), 135–159 (1952)

Dantzig, G.: Maximization of a linear function of variables subject to linear inequalities. In Koopmans, T.C. (ed.) Activity Analysis of Production and Allocation, pp. 101–111. Wiley, NY (1951). http://cowles.econ.yale.edu/P/cm/m13/m

Dantzig, G.: Reminiscences about the origins of linear programming. Oper. Res. Lett. **1**(2) (1982). http://www.sciencedirect.com/science/article/pii/0167637782900438

Hall, J.A.J., McKinnon, K.I.M.: The simplest examples where the simplex method cycles and conditions where EXPAND fails to prevent cycling. Math. Program. **100**, 133–150 (2000)

Karmarkar, N.: A new polynomial time algorithm for linear programming. Combinatorica **4**(4), 373–395 (1984). http://retis.sssup.it/~bini/teaching/optim2010/karmarkar.pdf

Koberstein, A., Suhl, U.H.: Progress in the dual simplex method for large scale LP problems: practical dual phase 1 algorithms. Comput. Optim. Appl. **37**, 49–65 (2007)

Mittelmann, H.D.: Test problems. http://plato.asu.edu/ftp/lp2.html. Accessed 25 Nov 2022

Nash, J.C.: The (Dantzig) simplex method for linear programming. Comput. Sci. Eng. **2**(1), 29–31 (2000)

Saff, E.B, Snider, A.D.: Fundamentals of Matrix Analysis with Applications (Sect. 1.6). Wiley, New York, (2016)

Nonlinear Programming in One Dimension

3

3.1 The Zero Derivative Rule and Its Limitations

The task of finding the extrema of a smooth function of one variable, $f(x)$, evokes cherished memories of our first calculus class. We'll assume that we are looking for a *minimum* point (to be specific), and that the cost function to be minimized is smooth enough to have two continuous derivatives (so we can use calculus).

The key observation is that if the function has a minimum at an interior point of an interval where its derivative is continuous, the derivative must be zero (Fig. 3.1).

This simple rule is confounded by a few anomalies:

1. *local extrema.* Some zero-slope points may only be *local* extrema, or *inflection points*; see Fig. 3.2. (These are sometimes called *critical points* in the literature.)
2. *non-smooth minima.* The minimum may occur at points where the function, or its derivative, are discontinuous (Fig. 3.3).
3. *end-point minimum.* If the domain of the function is restricted, the extrema may occur at the end points (Fig. 3.4).

If the formula for the derivative is easy to compute and its zeros can be calculated, one can simply evaluate, and tabulate, the values of f at these zeros. If necessary one can then compare these values with the end-point values and the values at nondifferentiable points, and select the maximum and minimum from this set. This is usually the easiest way to find extrema.

Example 3.1. *The easiest optimization problem of all* is to find the extreme point of the quadratic (parabola) $y = f(x) = ax^2 + bx + c$. The derivative is $2ax + b$, and it is zero when $x = -b/2a$. The extremal value of f equals $-b^2/4a + c$.

© The Author(s), under exclusive license to Springer Nature Switzerland AG 2023
A. D. Snider, *Basics of Optimization Theory*, Synthesis Lectures on Mathematics
& Statistics, https://doi.org/10.1007/978-3-031-29219-4_3

Fig. 3.1 Derivative equals
zero

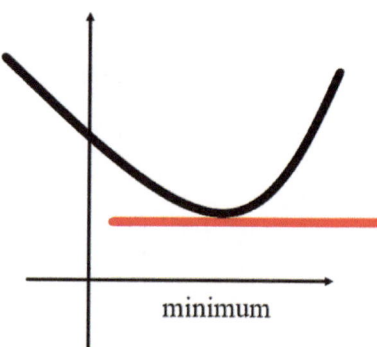

Fig. 3.2 Critical points

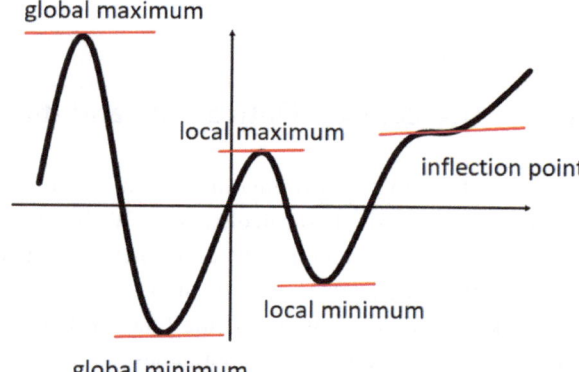

Fig. 3.3 Non-smooth minima

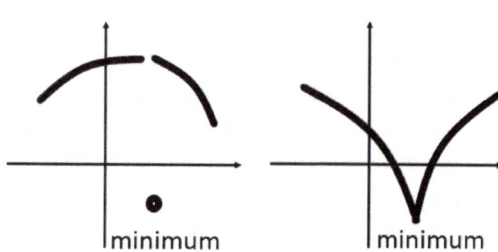

Fig. 3.4 End point extrema

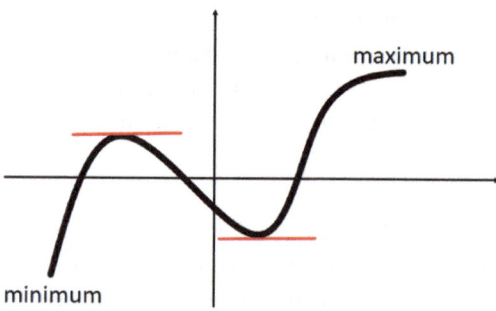

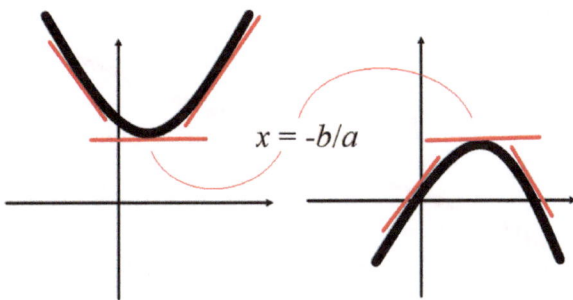

Fig. 3.5 $ax^2 + bx + c$ (**a**) $a > 0$ (**b**) $a < 0$

If the second derivative, $2a$, is positive, the first derivative is always rising, and the extremum is a minimum; see Fig. 3.5a. Otherwise the extremum is a maximum (Fig. 3.5b).

Example 3.2. Find the extrema of the function $\frac{\cos 3\pi x}{x}$ with $0.1 \leq x \leq 1.1$. (http://en.wik ipedia.org/wiki/Maxima_and_minima).

Answer. The zeros of the derivative satisfy

$$f'(x) = 0 = -3\pi \frac{\sin 3\pi x}{x} - \frac{\cos 3\pi x}{x^2}, \text{ or } 3\pi x = -\cot 3\pi x.$$

Graphs of the left and right hand sides of the latter equation are displayed in Fig. 3.6. The graphs cross near 0.297, 0.648, and 0.988.

By evaluating and comparing the values of $f(x)$ at the three zero-slope points and the end points 0.1 and 1.1, we find there is a global maximum at $x = 0.1$ (a boundary point), a global minimum near $x = 0.297$, a local maximum near $x = 0.648$, and a local minimum near $x = 0.988$ (Fig. 3.7).

3.2   Nonlinear Search

A good algorithm for finding the extrema of a function in one dimension, to high accuracy, is very important. In fact it is a key ingredient in *multidimensional* optimization. The problem with the straightforward procedure discussed in Sect. 3.1—finding the zeros of the derivative and comparing the corresponding values and the end-point values—may be confounded by either of two considerations:

(a) the calculation of the derivative may be too complicated to be practical; or it may even be impossible if, for instance, the formula for the function is unknown (so the function's values must be determined by measurement);

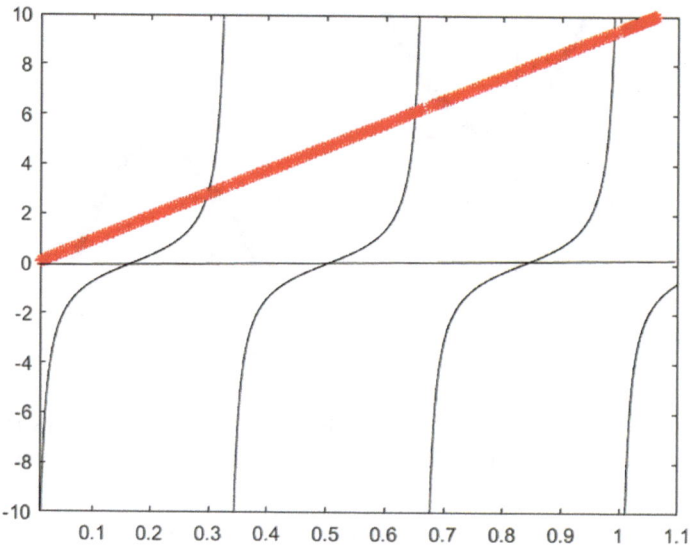

Fig. 3.6 $3\pi x$ (bold) and $-\cot 3\pi x$ (thin)

Fig. 3.7 Extrema of $\frac{\cos 3\pi x}{x}$ for $0.1 \leq x \leq 1.1$ "https://en. wikipedia.org/wiki/Maxima_ and_minima#/media/File:Ext rema_example_original.svg" is licensed under CC BY-SA 3.0 Unported license (https://creati vecommons.org/licenses/by-sa/ 3.0/legalcode)

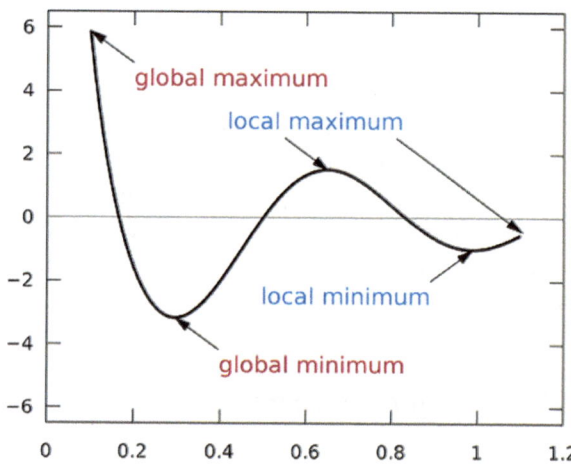

(b) even if the formula for the derivative can be calculated, there is no guarantee that we will be able to solve for its zeros. For example, it is known that, unlike the quadratic formula $(-b \pm \sqrt{b^2 - 4ac})/2a$, no formula exists for the roots of the quintic polynomial. And indeed, the derivative in Example 3.1 of the preceding section was a transcendental function, whose zeros we estimated from graphics.

So we will describe some *numerical* procedures for approximately locating a function's extrema—its minimum, to be specific. They are based on the same strategy: find a quadratic function $ax^2 + bx + c$ approximating the function to be optimized, calculate *its* extremum $-b/2a$, use it to redefine the quadratic approximation, and repeat.

Before we start, we need to have some idea of the shape of the graph of the cost function $f(x)$; we can only hope to approximate it, with parabolas, in subintervals where it is "bowl-shaped" (recall Fig. 3.7). One probably has to do some preliminary judicious sampling to get a rough feel for the location of the minimum.

So let's assume that by sampling you have found three points on the graph of $f(x)$, satisfying $x_1 < x_2 < x_3$ and $f(x_1) > f(x_2) < f(x_3)$, that lie on a bowl-shaped subregion containing the true minimum point (Fig. 3.8).

How do we choose a quadratic approximant $ax^2 + bx + c$ using these three data points? Since the approximant $ax^2 + bx + c$ has three undetermined constants, we can impose three constraints on it: we can either

(a) compute a, b, and c to force the approximant to pass through $f(x_1)$, $f(x_2)$, and $f(x_3)$, find its minimum point $-b/2a$, omit the point (x_1 or x_3) lying on the opposite side of x_2 from $-b/2a$, and start again using $-b/2a$ and the remaining two points;

(b) compute a, b, and c to force the approximant to pass through $f(x_2)$ and to match the slope of f at x_2 and at either x_1 or x_3—whichever has the lower value of f; then

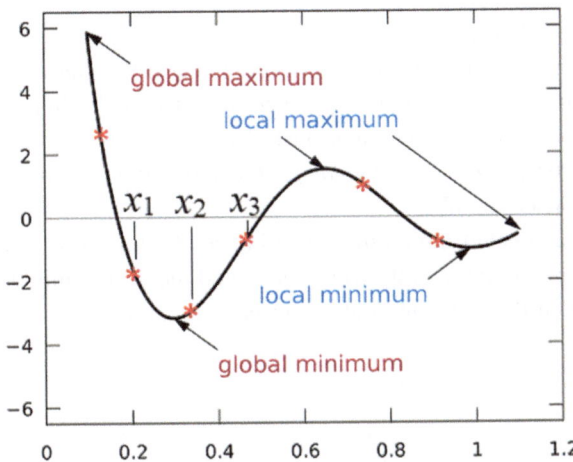

Fig. 3.8 Three sampled points near the global minimum "https://en.wikipedia.org/wiki/Max ima_and_minima#/media/File:Extrema_example_original.svg" is licensed under CC BY-SA 3.0 Unported license (https://creativecommons.org/licenses/by-sa/3.0/legalcode). Cropped and embellished from original

calculate $-b/2a$ and start over, matching the value of f at $-b/2a$ and the slopes of f at $-b/2a$ and x_2;

(c) compute $a, b,$ and c to force the approximant to pass through $f(x_2)$ and match the slope of f at x_2 and match the second derivative of f at x_2; then calculate $-b/2a$ and start over, matching the value, slope, and second derivative at $-b/2a$.

We omit the formulas for these calculations; they are well known and trivial to derive. Option (a) does not require the evaluation of any derivatives of f; option (b) requires the first derivative, and option (c) requires the second derivative also. The payoff for the extra work is expressed by the exponent in the following formula, which is valid when the iterates are sufficiently close to the true minimum:

[distance from the new estimated minimum to the true minimum] $\approx$

(constant) $\times$ *[distance from the previously estimated minimum to the true*

minimum]^s, where

$$s \approx 1.3 \text{ for option (a);}$$

$$s = \frac{2}{1 + \sqrt{5}} \approx 1.618 \text{ for option (b);}$$

$$s = 2 \text{ for option (c)}$$

(Luenberger and Ye (2008)).

A little calculation reveals that method (c) implements the classical Newton's method from calculus for finding the zeros of $f'(x)$, and that method (b) implements the classical "method of false position" for the same task. The constant $(1 + \sqrt{5})/2$ is familiar to Fibonacci enthusiasts; it is called the *golden ratio*, and is the asymptotic limit of the ratio of consecutive Fibonacci numbers.

Note that if the estimated minimum is within, say, 0.1 of its true value, the next estimate using option (c) will be roughly within 0.01, and the next within 0.0001. The number of *correct decimals* in the estimate doubles for each iteration.

Reference

Luenberger, D.G., Ye, Y.: Linear and Nonlinear Programming, 3rd edn. Springer Science & Business
  Media (2008)

Nonlinear Multidimensional Optimization

4

4.1   Visualizing the Objective Function

This chapter discusses techniques for locating the minimum (and maximum) of a nonlinear function of more than one variable, $f(x_1, x_2, \cdots, x_n)$. This means that the graph of the objective function is going to require at least three dimensions, and displaying it in print is a challenge.

Mapmakers devised a clever solution, at least for three dimensions. A "relief perspective" of a mountainous terrain (a ski area, in fact) is shown in Fig. 4.1.

The altitude, considered as a function of x and y coordinates, has its maxima at the mountaintops, of course. The mapmaker displays some curves connecting points at the same altitude: the *elevation contours* or *level curves* of the altitude function. If we abandon the effort to depict the mountain features but retain the elevation contours, we get the topographic map shown in Fig. 4.2.

It provides a reasonably clear, uncluttered visualization of the dynamics of the ski area. A mathematical depiction of the graphical process of plotting level curves is displayed in Fig. 4.3.

Note some features of the topographic map:

a. As we zero in, the elevation contours that surround most local maxima approximate a family of nested ellipses. The same is true for the local minima, of course. (For a ski area the valleys are usually much broader than the mountaintops).
b. In Figs. 4.1 and 4.2 a contour is drawn for every 20 feet of elevation. This enables us to see specific features of the terrain: if the contours are densely packed, a large change in elevation is achieved with only a small horizontal excursion—the terrain is *steep*. Sparsely spaced contours indicate flatter regions (Fig. 4.4).

© The Author(s), under exclusive license to Springer Nature Switzerland AG 2023   109
A. D. Snider, *Basics of Optimization Theory*, Synthesis Lectures on Mathematics
& Statistics, https://doi.org/10.1007/978-3-031-29219-4_4

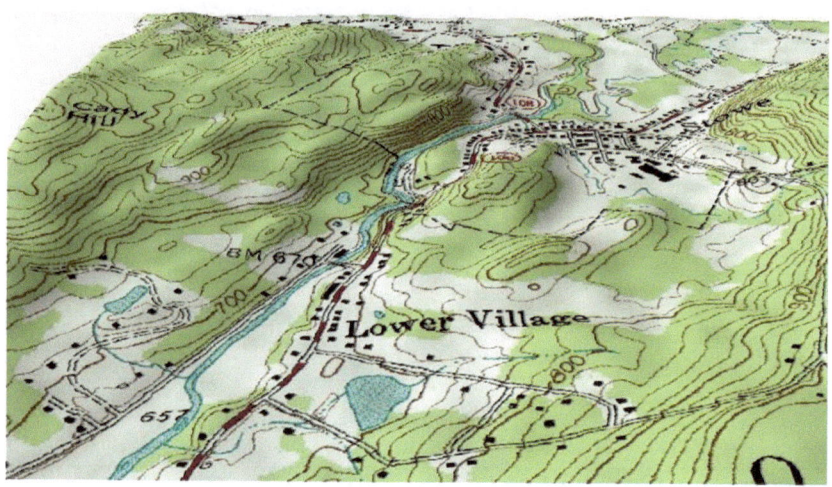

Fig. 4.1 "Topographic Relief perspective sample" by Kbh3rd at English Wikipedia (Aug. 2004) (https://en.wikipedia.org/wiki/File:Topographic-Relief-perspective-sample.jpg), used under CC BY-SA 2.0 Generic (https://creativecommons.org/licenses/by-sa/2.0/deed.en)

c. A skier is interested in getting downhill as fast as possible. (A skilled skier, anyway.) So he/she will head in the *direction of steepest descent*. Since any excursion *parallel to* an elevation contour is wasted, the direction of steepest descent is orthogonal to the corresponding elevation contour. The *grade* of a slope is a measure of its steepness, and the *gradient vector* encapsulates these details; by convention, it points in the direction of steepest *ascent*, and its length is proportional to the grade (Fig. 4.5). Note the *base* of the gradient vector is located at the site of interest, and its length indicates the steepness at that point; there is no significance to the point on the map lying under its tip.

Vector analysis tells us how to compute the gradient vector when the altitude can be expressed as a differentiable function of the (horizontal) coordinates $f(x, y)$. One simply assembles the vector of first-order derivatives:

$$\mathbf{grad}\, f(x, y) \equiv \nabla f = \frac{\partial f}{\partial x}\mathbf{i} + \frac{\partial f}{\partial y}\mathbf{j} = \left[\frac{\partial f}{\partial x} \; \frac{\partial f}{\partial y} \right]$$

It points in the direction of steepest ascent (highest rate of increase) of f, and has magnitude equal to the grade (i.e., this highest rate). In n dimensions the analogous vector of partial derivatives has the same properties. (In fact, the rate of increase of f in *any* direction specified by the unit vector $\mathbf{u}$ is given by the dot product $\nabla f \cdot \mathbf{u}$) (Davis and Snider 1995).

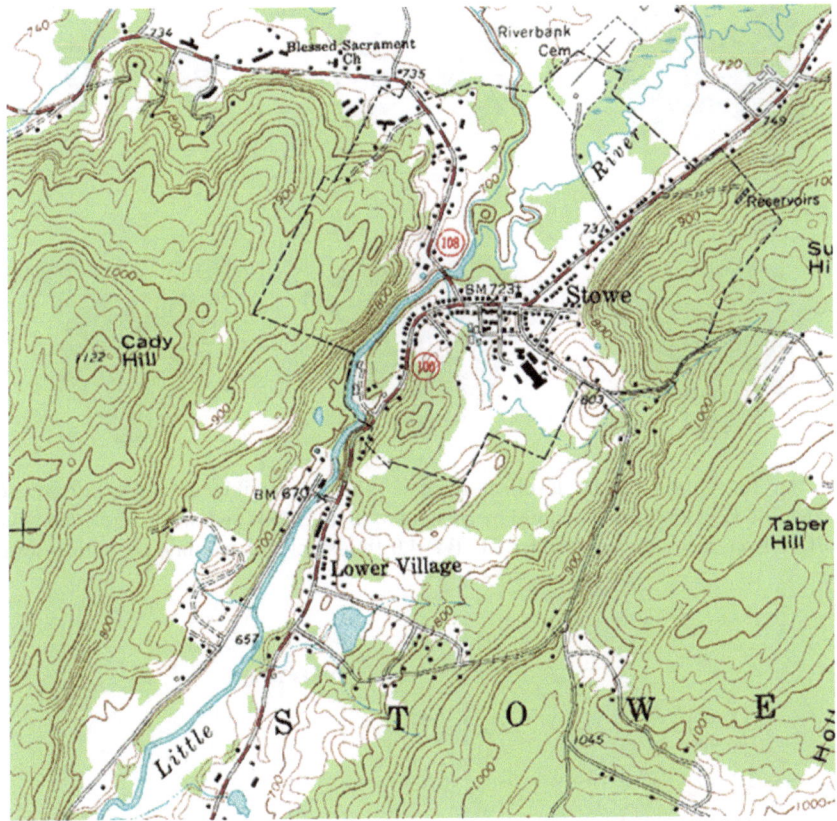

Fig. 4.2 "A topographic map of Stowe, Vermont with contour lines" (https://commons.wikime dia.org/wiki/File:Topographic_map_example.png) by USGS (Sept. 2005) (USGS Digital Raster Graphic file o44072d6.tif) is in the public domain

Important properties of the gradient

1. **grad** $f(x, y) \equiv \nabla f = \frac{\partial f}{\partial x}\mathbf{i} + \frac{\partial f}{\partial y}\mathbf{j} = \left[\frac{\partial f}{\partial x} \; \frac{\partial f}{\partial y}\right]$
2. **grad** f points in steepest ascent direction
3. $|$**grad** $f| =$ rate of increase in steepest ascent direction
4. **grad** f is perpendicular to contour $f =$ constant
5. **grad** f points directly *out of* region $f \leq$ constant.

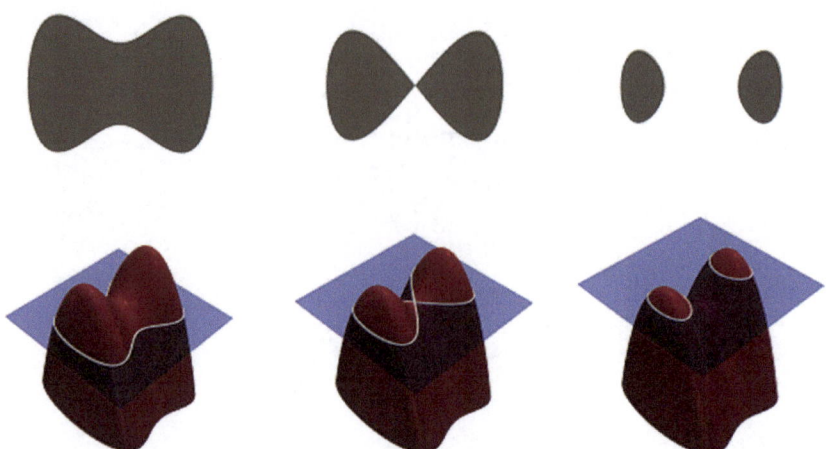

Fig. 4.3 "An Illustration of the level-set method" (https://en.wikipedia.org/wiki/Level-set_met hod#/media/File:Level_set_method.png) by Nicoguaro (Feb. 2018) (own work, https://en.wikipedia. org/wiki/Level-set_method), used under CC BY 4.0 (https://creativecommons.org/licenses/by/4.0/ deed.en)

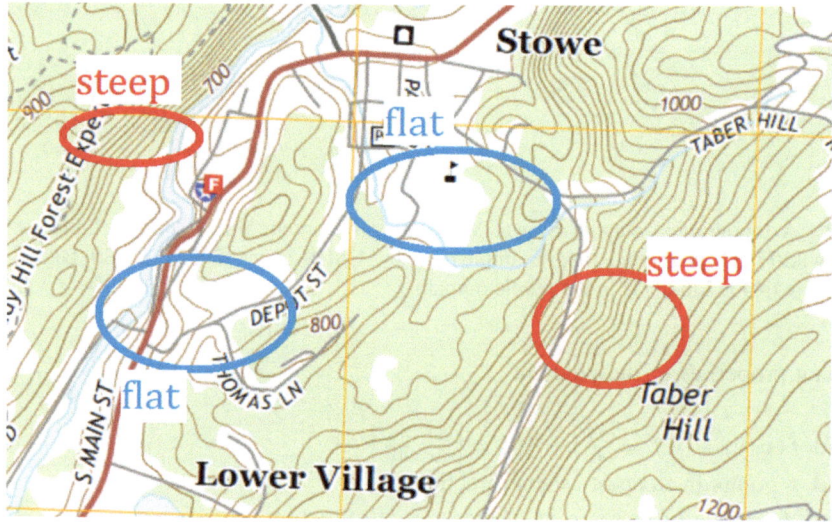

Fig. 4.4 "Contour map of Stowe, Vermont" (https://prd-tnm.s3.amazonaws.com/StagedProducts/ Maps/USTopo/PDF/VT/VT_Stowe_20150706_TM_geo.pdf) by USGS (U.S. Geological Survey; Department of the Interior/USGS) is in the public domain. Cropped and embellished from original

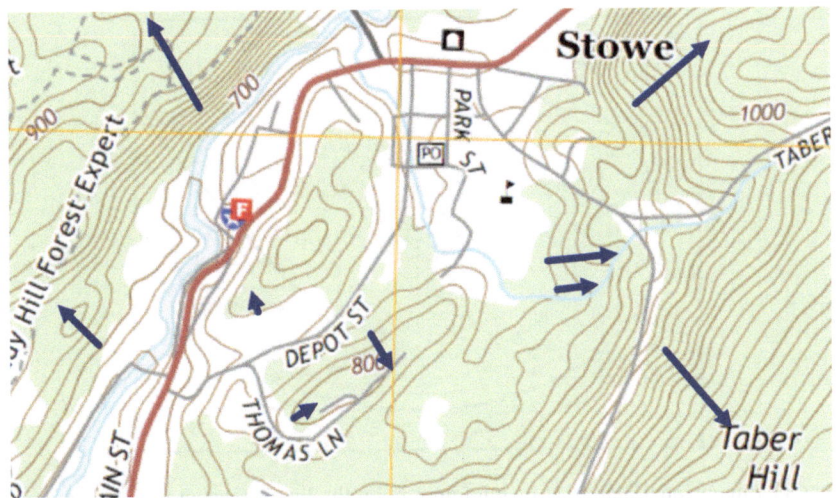

Fig. 4.5 (Steepest ascent) Gradient vectors (https://prd-tnm.s3.amazonaws.com/StagedProducts/Maps/USTopo/PDF/VT/VT_Stowe_20150706_TM_geo.pdf) by USGS (U.S. Geological Survey; Department of the Interior/USGS) is in the public domain. Cropped and embellished from original

4.2 Multidimensional Search

In this section we are going to study search methods for finding the *minimum* of a nonlinear function. Starting from any point on the contour map, we begin by choosing a direction and finding the function's minimum point on the line pointing in that direction. This is a one-dimensional search, and we can apply the techniques described in the preceding chapter.

It is important to have a mental picture of this process. The search is depicted in Fig. 4.6 for a function with very simple contours. Starting from the initial point on the outer contour, the function values diminish until we get to the black contour, then they increase.

Note that the search line will be *tangent* to the level curve passing through its minimum point. The analogous property holds in higher dimensions (Fig. 4.7).

Starting from the one-dimensional minimum located in Fig. 4.6, we choose another direction and search again; see Fig. 4.8.

And so on (Fig. 4.9). Presumably we will converge to the true minimum point.

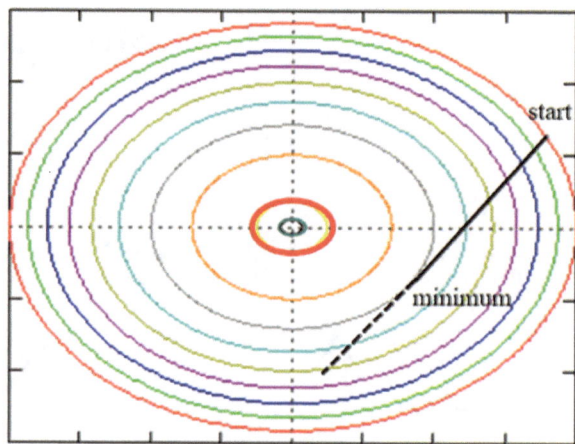

Fig. 4.6 One dimensional search: at its minimum point the line is tangent to the local contour

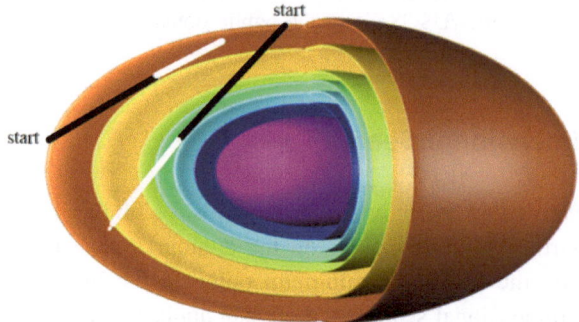

Fig. 4.7 The line and the local surface contour are tangent at the minimal point on the line.
(https://pixabay.com/el/illustrations/%CF%83%CF%86%CE%B1%CE%AF%CF%81%CE%B1-%
CE%BF%CE%BC%CF%8C%CE%BA%CE%B5%CE%BD%CF%84%CF%81%CE%BF%CF%
82-3d-5223299/) by caeuje is licensed by Pixabay License (https://pixabay.com/el/service/license/).
Stretched and embellished from original

4.3   Mathematical Characteristics of the Objective Function

The objective function to be minimized $f(x, y)$ is said to be *quadratic* if it takes the form

$$f(x, y) = a_{11}x^2 + a_{22}y^2 + 2a_{12}xy + b_1x + b_2y + c. \qquad (4.1)$$

If you have never performed this calculation, you should verify the following identity:

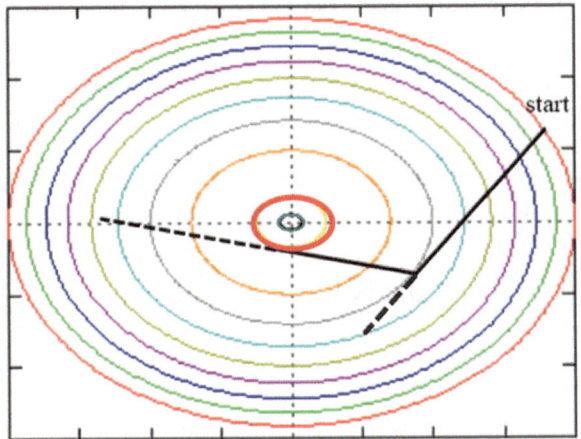

Fig. 4.8 Continuation of line search from Fig. 4.6

Fig. 4.9 Continuation of line search from Fig. 4.8

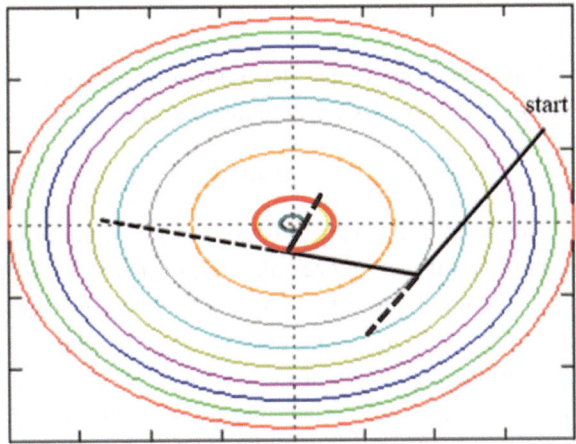

$$[x \; y] \begin{bmatrix} a_{11} \; a_{12} \\ a_{12} \; a_{22} \end{bmatrix} \begin{bmatrix} x \\ y \end{bmatrix} + [b_1 \; b_2] \begin{bmatrix} x \\ y \end{bmatrix} + c \equiv x^T A x + b^T x + c$$

$$= a_{11}x^2 + a_{22}y^2 + 2a_{12}xy + b_1 x + b_2 y + c \qquad (4.2)$$

where x is the column vector $\begin{bmatrix} x \\ y \end{bmatrix}$, A is the *symmetric*(!) matrix $\begin{bmatrix} a_{11} \; a_{12} \\ a_{12} \; a_{22} \end{bmatrix}$, and b is

the column vector $\begin{bmatrix} b_1 \\ b_2 \end{bmatrix}$. Observe that (4.2) agrees with (4.1). You will see why A can always be taken to be symmetric, and where the factor "2" in (4.1) comes from.

(The matrix $A = \begin{bmatrix} a_{11} & 6 \\ 4 & a_{22} \end{bmatrix}$ gives the same quadratic form as $\begin{bmatrix} a_{11} & 5 \\ 5 & a_{22} \end{bmatrix}$.)

Now verify that $\frac{\partial \mathbf{b}^T \mathbf{x}}{\partial x} = b_1$ and $\frac{\partial \mathbf{b}^T \mathbf{x}}{\partial y} = b_2$, so that the components of the partials of $\mathbf{b}^T \mathbf{x}$ can be arranged into a column vector validating the identity

$$\mathbf{grad}\ \mathbf{b}^T \mathbf{x} = \mathbf{b} = \begin{bmatrix} \frac{\partial}{\partial x} \\ \frac{\partial}{\partial y} \end{bmatrix} (\mathbf{b}^T \mathbf{x}) \tag{4.3}$$

Next verify that $\frac{\partial \mathbf{x}^T \mathbf{A} \mathbf{x}}{\partial x} = 2a_{11}x + 2a_{12}y$ and $\frac{\partial \mathbf{x}^T \mathbf{A} \mathbf{x}}{\partial y} = 2a_{12}x + 2a_{22}y$, so that the components of the partials of $\mathbf{x}^T \mathbf{A} \mathbf{x}$ can be arranged into a column vector validating the identity

$$\mathbf{grad}\ \mathbf{x}^T \mathbf{A} \mathbf{x} = 2\mathbf{A}\mathbf{x} \quad \left(= \begin{bmatrix} \frac{\partial}{\partial x} \\ \frac{\partial}{\partial y} \end{bmatrix} (\mathbf{x}^T \mathbf{A} \mathbf{x}) \right). \tag{4.4}$$

It is easy to show that the n-dimensional quadratic form $\mathbf{x}^T \mathbf{A} \mathbf{x} + \mathbf{b}^T \mathbf{x} + c$ that results when $\mathbf{x}$ and $\mathbf{b}$ are n-by-1 columns and $\mathbf{A}$ is an n-by-n matrix satisfies the same identities:

$$\mathbf{grad}(\mathbf{x}^T \mathbf{A} \mathbf{x} + \mathbf{b}^T \mathbf{x} + c) = 2\mathbf{A}\mathbf{x} + \mathbf{b}. \tag{4.5}$$

Now if $\mathbf{x}^{min}$ happens to be a local minimum point of an objective function (quadratic or not), and if we "freeze" all components of $\mathbf{x}$ except one (say, the jth), then obviously x_j^{min} is a minimum for the resulting *univariate* objective function; and therefore, by the reasoning of Chap. 3, the derivative of the objective with respect to x_j is zero at $\mathbf{x}^{min}$ (barring the anomalies discussed there). But this is precisely the partial derivative with respect to x_j at $\mathbf{x}^{min}$. Therefore *all first-order partial derivatives of a (smooth) objective function are zero at its minimum points.*

In particular, if the objective function is quadratic, its minima satisfy

$$\mathbf{grad}(\mathbf{x}^T \mathbf{A} \mathbf{x} + \mathbf{b}^T \mathbf{x} + c) = 2\mathbf{A}\mathbf{x} + \mathbf{b} = \mathbf{0} \text{ (the zero column vector)},$$

and if A is invertible then

$$\mathbf{x}^{min} = -\mathbf{A}^{-1}\mathbf{b}/2. \tag{4.6}$$

Equation (4.6) is the generalization of $x^{min} = -\frac{b}{2a}$ for the minimum point on a parabola. It is only a *necessary* condition for a minimum. It shows that *if* a quadratic objective has local minima and $\mathbf{A}$ is invertible, then (4.6) holds (and thus the minimum is unique).

To derive a *sufficient* condition, observe the surprising identity (you can almost work it out mentally)

$$\mathbf{x}^T\mathbf{A}\mathbf{x} + \mathbf{b}^T\mathbf{x} + c = \left(\mathbf{x} + \mathbf{A}^{-1}\mathbf{b}/2\right)^T\mathbf{A}(\mathbf{x} + \mathbf{A}^{-1}\mathbf{b}/2) - \mathbf{b}^T\mathbf{A}^{-1}\mathbf{b}/4 + c$$

$$= \left(\mathbf{x} - \mathbf{x}^{min}\right)^T\mathbf{A}\left(\mathbf{x} - \mathbf{x}^{min}\right) - \mathbf{b}^T\mathbf{A}^{-1}\mathbf{b}/4 + c$$

$$= \mathbf{y}^T\mathbf{A}\mathbf{y} + constant$$

where $\mathbf{y} = \mathbf{x} - \mathbf{x}^{min}$ and the constant is $-\mathbf{b}^T\mathbf{A}^{-1}\mathbf{b}/4 + c$. Now in matrix theory a symmetric matrix is said to be *positive definite* if the number $\mathbf{y}^T\mathbf{A}\mathbf{y}$ is positive for every value of $\mathbf{y}$ except $\mathbf{0}$.

To interpret the positive-definiteness condition, note that $\mathbf{y}^T\mathbf{A}\mathbf{y}$ is the dot product of the vectors $\mathbf{y}$ and $\mathbf{A}\mathbf{y}$, and as such is proportional to the cosine of the angle between them. Therefore $\mathbf{A}$ is positive definite if $\mathbf{A}\mathbf{y}$ is less than 90 degrees away from $\mathbf{y}$, for every nonzero vector $\mathbf{y}$. Many conditions guaranteeing positive definiteness are well-known. Most relevant to us here: if all the eigenvalues of $\mathbf{A}$ are positive then $\mathbf{A}$ is positive definite.

So if $\mathbf{A}$ is positive definite, the quadratic objective function $\mathbf{x}^T\mathbf{A}\mathbf{x} + \mathbf{b}^T\mathbf{x} + c$ is never less than its value $(-\mathbf{b}^T\mathbf{A}^{-1}\mathbf{b}/4 + c)$ at $\mathbf{x}^{min} = -\mathbf{A}^{-1}\mathbf{b}/2$, which consequently is its minimum point.

To simplify the discussion of what follows in this chapter we are going to be somewhat cavalier and make some claims that are rigorously true only for quadratic objective functions. We defend this heinous crime with the following argument.

The Taylor expansion for an arbitrary, smooth two-variable function around a point x_0, y_0 can be derived by first freezing y and expanding the one-dimensional function of x:

$$f(x, y) \approx f(x_0, y) + \frac{\partial f(x_0, y)}{\partial x}(x - x_0) + \frac{1}{2!}\frac{\partial^2 f(x_0, y)}{\partial x^2}(x - x_0)^2$$
$$+ \text{ higher order terms};$$

then we Taylor-expand the y variation in the coefficients

$$f(x, y) \approx f(x_0, y) + \frac{\partial f(x_0, y)}{\partial x}(x - x_0) + \frac{1}{2!}\frac{\partial^2 f(x_0, y)}{\partial x^2}(x - x_0)^2$$

$$\| \qquad\qquad \| \qquad\qquad\qquad \|$$

$$f(x_0, y_0) + \frac{\partial f(x_0, y_0)}{\partial x}(x - x_0) + \frac{1}{2!}\frac{\partial^2 f(x_0, y_0)}{\partial x^2}(x - x_0)^2$$

$$+ \qquad\qquad\qquad +$$

$$\frac{\partial f(x_0, y_0)}{\partial y}(y - y_0) + \frac{\partial^2 f(x_0, y_0)}{\partial y \partial x}(y - y_0)(x - x_0)$$

$$+$$

$$\frac{1}{2!}\frac{\partial^2 f(x_0, y_0)}{\partial y^2}(y - y_0)^2 \qquad + \qquad \text{higher order terms}$$

So if (x, y) is close enough to (x_0, y_0) that we can neglect the higher order terms, we can write

$$f(x, y) \approx \frac{1}{2}\begin{bmatrix} x - x_0 \\ y - y_0 \end{bmatrix}^T \begin{bmatrix} \frac{\partial^2 f(x_0, y_0)}{\partial x^2} & \frac{\partial^2 f(x_0, y_0)}{\partial y \partial x} \\ \frac{\partial^2 f(x_0, y_0)}{\partial y \partial x} & \frac{\partial^2 f(x_0, y_0)}{\partial y^2} \end{bmatrix} \begin{bmatrix} x - x_0 \\ y - y_0 \end{bmatrix}$$

$$+ \begin{bmatrix} \frac{\partial f(x_0, y_0)}{\partial x} \\ \frac{\partial f(x_0, y_0)}{\partial y} \end{bmatrix}^T \begin{bmatrix} (x - x_0) \\ (y - y_0) \end{bmatrix} + f(x_0, y_0)$$

$$\approx (\mathbf{x} - \mathbf{x_0})^T \mathbf{A}(\mathbf{x} - \mathbf{x_0}) + \mathbf{b}^T(\mathbf{x} - \mathbf{x_0}) + f(x_0, y_0)$$

$$\approx \mathbf{x}^T \mathbf{A}\mathbf{x} + (\mathbf{b} - 2\mathbf{A}\mathbf{x_0})^T \mathbf{x} + \mathbf{x_0}^T \mathbf{A}\mathbf{x} + f(x_0, y_0)$$

$$\equiv \mathbf{x}^T \mathbf{A}\mathbf{x} + \tilde{\mathbf{b}}^T \mathbf{x} + c.$$

Thus near any point, f can be *approximated* by a quadratic function. The column vector $\tilde{\mathbf{b}}$ is the gradient and the matrix A is (one-half of) the *Hessian*—the matrix of second partials—evaluated at the expansion point (x_0, y_0).

The approximation can obviously be generalized to n dimensions, and we are justified in serendipitously supposing that smooth nonlinear objective functions with positive definitive Hessians behave like quadratics in sufficiently small neighborhoods; their level surfaces are approximately ellipsoidal. However we shall restrict our discussion to two dimensions (Fig. 4.10).

Fig. 4.10 (Approximate) level curves for a smooth nonlinear objective function

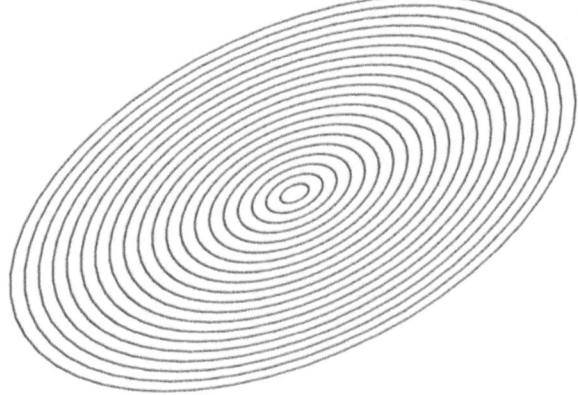

Fig. 4.11 Coordinate Descent

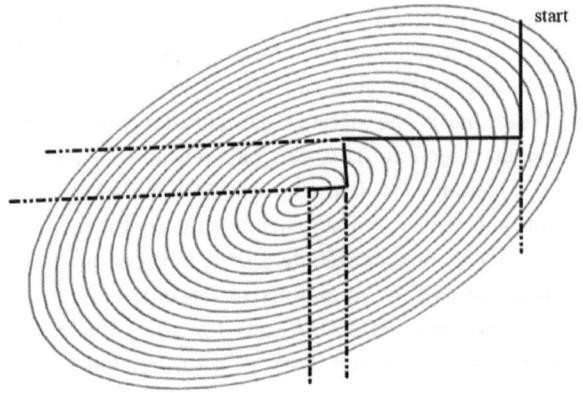

4.4 Coordinate Descent

Coordinate descent is the simplest strategy for choosing search directions; one first minimizes $f(x_1, x_2, \cdots, x_n)$ over x_1, holding x_2 through x_n constant; then over x_2, over x_3, etc. Naturally an algorithm this simple has to have a downside; it is typically slow to converge. See Fig. 4.11.

4.5 Method of Steepest Descent

The directions of steepest descent (*negative* gradients) starting from the outer contour are normal to that contour; they are displayed in Fig. 4.12.

They each provide the best local minimizer directions *at the start*, but unless the level contours are circles only four starting locations (the apogees and perigees of the level

Fig. 4.12 Steepest descent
directions

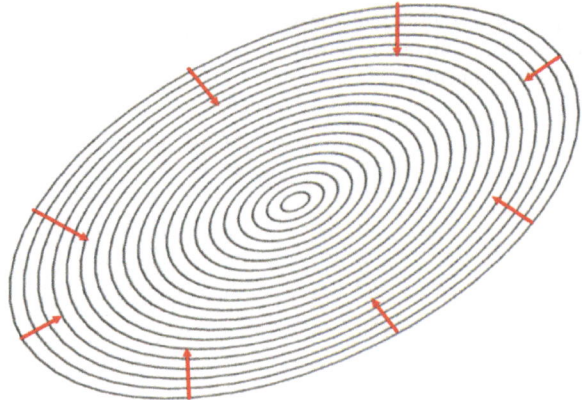

surfaces) will get you to the minimum on the first try. Generally, conducting a series of
one-dimensional searches along steepest descent directions leads to a meandering path
converging to the minimum (Fig. 4.13).

Note that in the steepest descent algorithm each search direction is orthogonal to the
previous direction. This can be seen as follows: at the minimum point of the function
$f(x, y)$ along *any* line, the line will be tangent to the level curve of f through that
point; and the subsequent steepest descent direction there will be orthogonal to the level
curve. (Of course consecutive search directions for coordinate search (Sect. 4.4) are also
orthogonal, but for an entirely different reason.)

Various successful strategies for straightening out the unproductive meanderings of
steepest descent search have been formulated. After computing the gradient, these techniques try to null out gradient components that have exhibited a history of back-and-forth
oscillation, thus selecting *deflected gradient* directions for the line searches (Gecan and
Snider 1995).

Fig. 4.13 Steepest descent
search

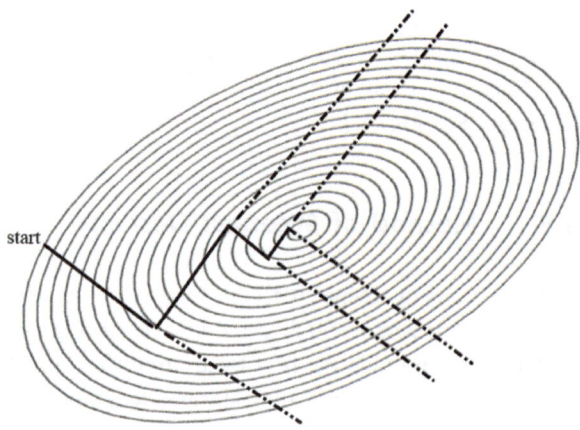

4.6   Conjugate Directions

There is one type of objective function whose minimization is achieved very quickly by coordinate and steepest descent searches: namely, functions whose level curves are concentric circles. Consider Fig. 4.14a.

A *steepest descent* search would land on the minimum point after a single line search (Fig. 4.14b). *Coordinate* search would find the minimum after *two* line searches (Fig. 4.14c); indeed, two consecutive line searches along *any* two mutually orthogonal search directions would suffice (Fig. 4.14d). In three (n) dimensions, convergence is achieved for concentric spherical level surfaces after searching along any three (n) mutually orthogonal directions (Fig. 4.15a)—or, after one steepest descent search (Fig. 4.15b). (If Fig. 4.15a is too obscure, imagine an unskilled sniper trying to hit a target. Each

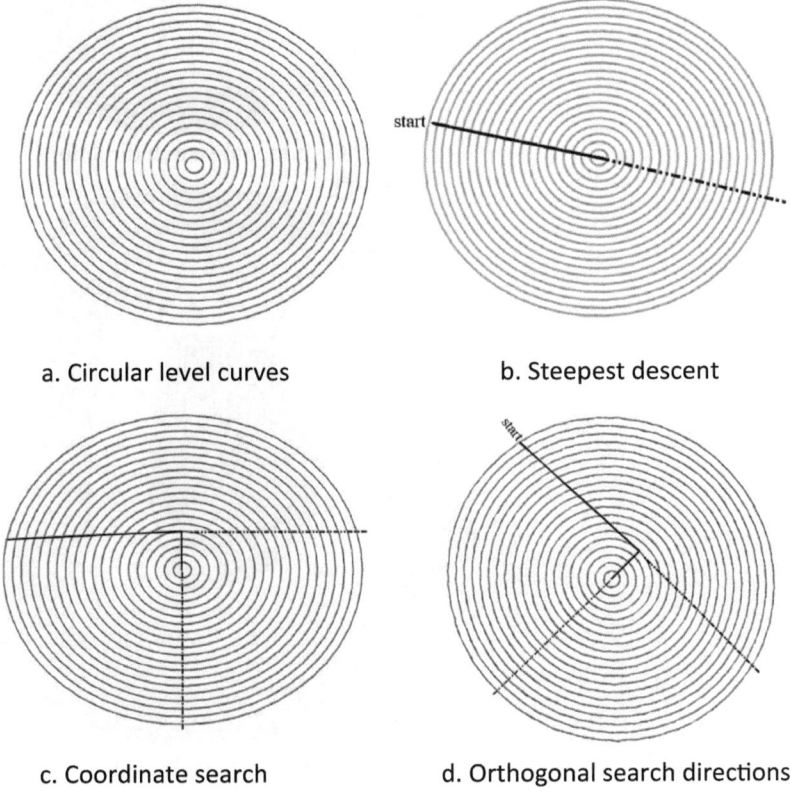

a. Circular level curves   b. Steepest descent

c. Coordinate search   d. Orthogonal search directions

Fig. 4.14 Searches along circular level curves

Fig. 4.15 Orthogonal and steepest descent search for spherical level surfaces (https://pixabay.com/el/illustrations/%CF%83%CF%86%CE%B1%CE%AF%CF%81%CE%B1-%CE%BF%CE%BC%CF%8C%CE%BA%CE%B5%CE%BD%CF%84%CF%81%CE%BF%CF%82-3d-5223299/) by caeuje is licensed by Pixabay License (https://pixabay.com/el/service/license/). Stretched and embellished from original

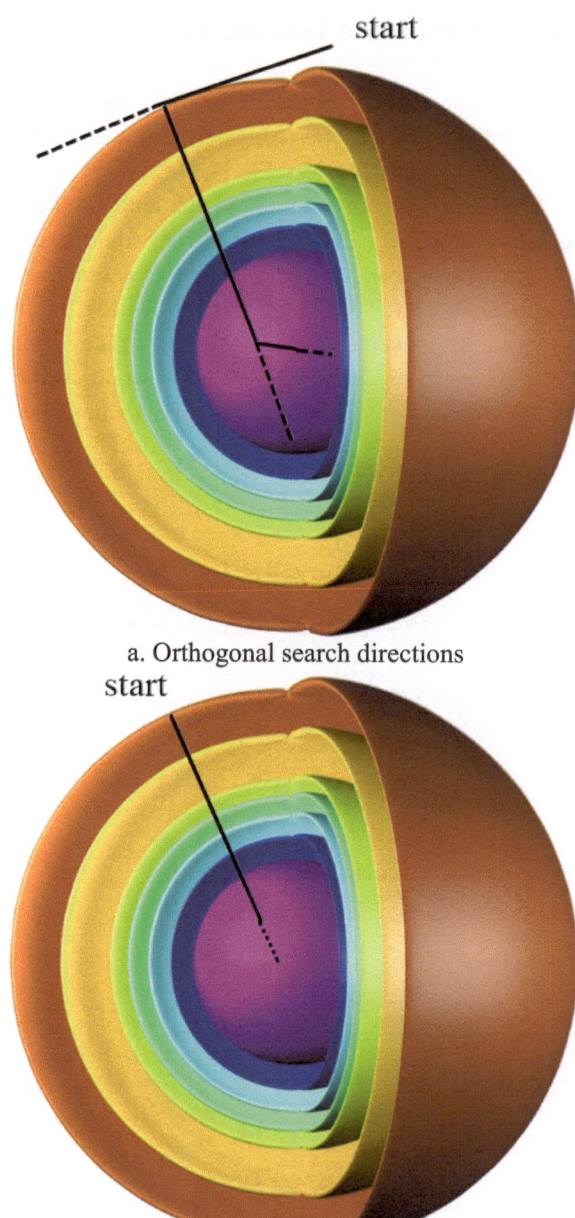

a. Orthogonal search directions

b. Steepest descent

time he misses, he gets another shot from the point of closest approach along the preceding trajectory. Then he doesn't need to aim; by simply ensuring that each shot is fired
perpendicular to the previous shots, he is guaranteed to hit the target on the third shot.)

These considerations lend us to speculate whether the circular/spherical geometry
advantages can be exploited for other objective functions.

What we would like to have is a change of coordinates such that in the new coordinate system the level curves are concentric circles. Consider, for example, the objective
function $f(x, y) = 4x^2 + y^2$. Its level curves are perfect ellipses; the curve $4x^2 + y^2 = 4$,
for instance, cuts the x axis at $x = \pm 1$, and the y axis at $y = \pm 2$ (Fig. 4.16a). But if
we change variables and let $\xi = 2x$, we stretch the horizontal axis like an accordion and
the level curves of the same objective, $f(\xi/2, y) = 4\left(\frac{\xi}{2}\right)^2 + y^2 = \xi^2 + y^2$, appear as
concentric circles (Fig. 4.16b).

Thus we could minimize f in one step with steepest descent or 2 steps of orthogonal
direction search in ξ, y space.

The catch, as you might have guessed, is that it can be very difficult to find the
change of variables that transforms the level curves (surfaces) of f into concentric circles
(spheres). Failing that, we might hope, at least, to be able to do one of the following.

1. Find the path, in the original-variable-space, that corresponds to the line of steepest
   descent in the new variables. A one-dimensional search along this path would locate
   the minimum in one step.

   Well, this is unrealistic. It's almost as difficult as finding the change of variables
   itself. We cannot implement this strategy.
2. Find a way to choose directions, in the original-variable-space, that correspond to
   orthogonal directions in the new variables. *Conjugate directions* is the common
   nomenclature for directions in the original space corresponding to orthogonal directions in the new space. See Fig. 4.17. For an n-dimensional quadratic objective
   function, line searches along mutually conjugate directions would locate the minimum
   in n steps.

Finding conjugate directions is not nearly as exhausting as finding the change of variables.
For quadratic objective functions it can be readily implemented, yielding the predicted
convergence if exact calculations are performed, and yielding very good performance
otherwise.

We defer the computational details of implementing conjugate direction search to the
next section. Presuming we have this facility at hand, let us reconsider the task of locating
the minimum. Although we have reduced the task from an infinite number of searches
(Fig. 4.13) to a finite number (Fig. 4.14), if the problem at hand has, say, thousands of
variables, we are still faced with the prospect of performing thousands of searches! Is
there a way we can choose the conjugate directions to get us *close* to the minimum in a
shorter time? The *conjugate gradient* algorithm accomplishes this.

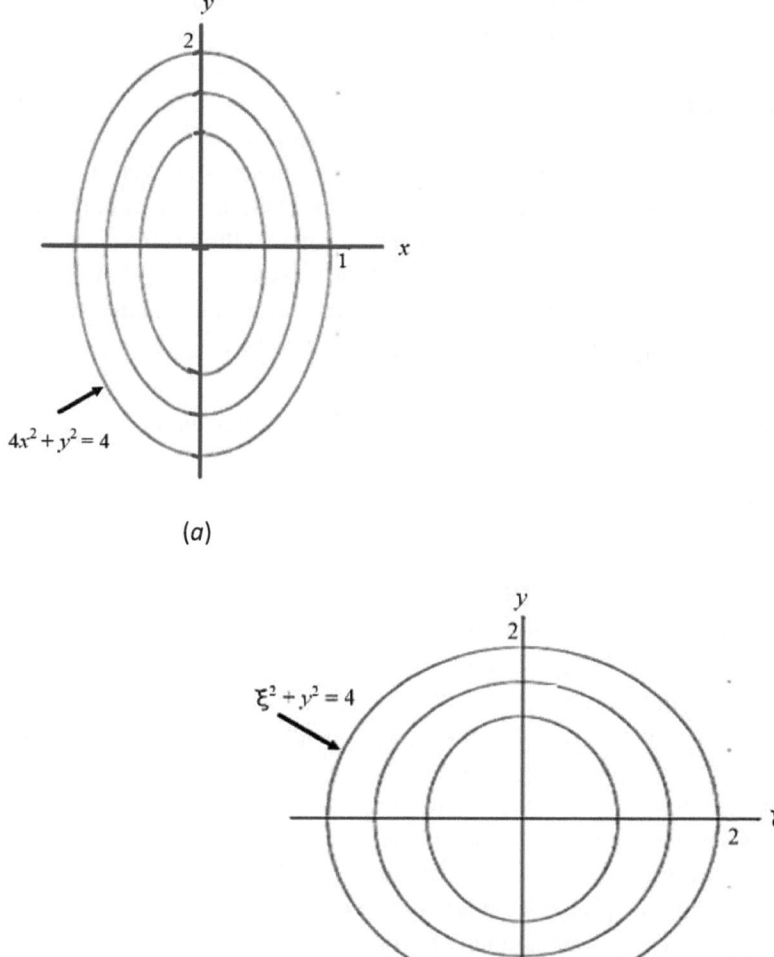

(a)

(b)

Fig. 4.16 The level curves of $4x^2 + y^2$ become circles under the mapping $\xi = 2x$

The best direction to start a search, *at least in the short term*, is the direction of steepest descent—the negative of the gradient (computed in the old variables). So we perform one line search along steepest descent. Then we reevaluate the gradient; but before we begin the next search, we subtract off a multiple of the old gradient so that the resulting search direction is conjugate to the previous direction.

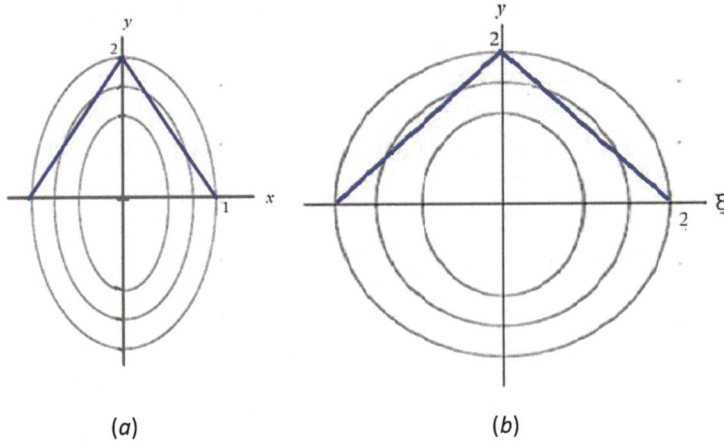

(a)   *(b)*

Fig. 4.17   Conjugate directions map to orthogonal directions

Next we perform the line search, and reevaluate the gradient again, and subtract off multiples of the previous search directions to find a new direction that is conjugate to all the previous directions. And so on.

So after n steps, we will have conducted n searches along mutually conjugate directions, and we will arrive at the minimum—guaranteed! (Well, for quadratic positive definite objective functions.) But since our choice of directions has been "guided" by steepest descents (though not dictated by them), we are likely to approach the minimum much faster; and probably we can terminate the algorithm with fewer steps, achieving satisfactory results.

That's the conjugate gradient algorithm. As you will see in the next section, it is surprising easy to code, for such a sophisticated notion. In practice it performs very well, particularly when it is tweaked with adjustments to correct for inaccurate numerics and multiple minimum points. (Golub and O'Leary 1989).[1]

4.7   Details (Optional Reading): The Conjugate Gradient Algorithm

Assume we seek to minimize a quadratic objective function $f(\mathbf{x}) = \mathbf{x}^T\mathbf{A}\mathbf{x} + \mathbf{b}^T\mathbf{x} + c$ whose Hessian, $2\mathbf{A}$, is positive definite (and thus so is $\mathbf{A}$). We propose to display a linear change of coordinates that renders the level surfaces of f as concentric circles, formulate

---

[1] Commercial software bearing the names of W. C. Davidon, R. Fletcher, M. J. D. Powell, C. M. Reeves, and J. K. Reid is readily available. The MATLAB® Optimization Toolbox™ is anchored by an algorithm amalgamized from the works of Broyden, C. G., Fletcher, R., Goldfarb, D., and Shanno, D. F.; see Lam (2020).

the notion of conjugate directions for f, and demonstrate that the computation of conjugate directions is easy.

The first step of the *analysis* is formidable. Ultimately we'll have to show how to bypass it in practice. Presuming that $\mathbf{A}$ is symmetric, we consider its diagonalization by an orthogonal matrix $\mathbf{P}$. Recall that the eigenvalues λ_i are all positive, and write

$$
\mathbf{A} = \mathbf{P}^T \Lambda \mathbf{P} =
\begin{bmatrix} \cdots \mathbf{P} \cdots \end{bmatrix}^T
\begin{bmatrix}
\lambda_1 & 0 & \cdots & 0 \\
0 & \lambda_2 & \cdots & 0 \\
\vdots & \vdots & \ddots & \vdots \\
0 & 0 & \cdots & \lambda_n
\end{bmatrix}
\begin{bmatrix} \cdots \mathbf{P} \cdots \end{bmatrix}
$$

$$
= \underbrace{
\begin{bmatrix} \cdots \mathbf{P} \cdots \end{bmatrix}^T
\begin{bmatrix}
\lambda_1^{1/2} & 0 & \cdots & 0 \\
0 & \lambda_2^{1/2} & \cdots & 0 \\
\vdots & \vdots & \ddots & \vdots \\
0 & 0 & \cdots & \lambda_n^{1/2}
\end{bmatrix}}_{\mathbf{S}^T}
\underbrace{
\begin{bmatrix}
\lambda_1^{1/2} & 0 & \cdots & 0 \\
0 & \lambda_2^{1/2} & \cdots & 0 \\
\vdots & \vdots & \ddots & \vdots \\
0 & 0 & \cdots & \lambda_n^{1/2}
\end{bmatrix}
\begin{bmatrix} \cdots \mathbf{P} \cdots \end{bmatrix}}_{\mathbf{S}}
$$

$$
= \mathbf{S}^T \mathbf{S}.
$$

Now execute the change of variables $\mathbf{z} = \mathbf{S}\mathbf{x}$, $\mathbf{x} = \mathbf{S}^{-1}\mathbf{z}$. Note that

$$
\mathbf{S}^{-1} =
\begin{bmatrix} \cdots \mathbf{P} \cdots \end{bmatrix}^T
\begin{bmatrix}
\lambda_1^{-1/2} & 0 & \cdots & 0 \\
0 & \lambda_2^{-1/2} & \cdots & 0 \\
\vdots & \vdots & \ddots & \vdots \\
0 & 0 & \cdots & \lambda_n^{-1/2}
\end{bmatrix}.
$$

Thus the objective function, in the z variables, contains a lot of identity $(\mathbf{P}\mathbf{P}^T)$ matrices:

$$
\hat{f} = (\mathbf{S}^{-1}\mathbf{z})^T \mathbf{A}(\mathbf{S}^{-1}\mathbf{z}) + \mathbf{b}^T(\mathbf{S}^{-1}\mathbf{z}) + c = \mathbf{z}^T(\mathbf{S}^{-1})^T \mathbf{A}\mathbf{S}^{-1}\mathbf{z} + \mathbf{b}^T\mathbf{S}^{-1}\mathbf{z} + c
$$

$$
= \mathbf{z}^T
\begin{bmatrix}
\lambda_1^{-\frac{1}{2}} & 0 & \cdots & 0 \\
0 & \lambda_2^{-\frac{1}{2}} & \cdots & 0 \\
\vdots & \vdots & \ddots & \vdots \\
0 & 0 & \cdots & \lambda_n^{-\frac{1}{2}}
\end{bmatrix}
\mathbf{P}\mathbf{P}^T
\begin{bmatrix}
\lambda_1 & 0 & \cdots & 0 \\
0 & \lambda_2 & \cdots & 0 \\
\vdots & \vdots & \ddots & \vdots \\
0 & 0 & \cdots & \lambda_n
\end{bmatrix}
\mathbf{P}
$$

$$
\cdot \mathbf{P}^T
\begin{bmatrix}
\lambda_1^{-\frac{1}{2}} & 0 & \cdots & 0 \\
0 & \lambda_2^{-\frac{1}{2}} & \cdots & 0 \\
\vdots & \vdots & \ddots & \vdots \\
0 & 0 & \cdots & \lambda_n^{-\frac{1}{2}}
\end{bmatrix}
\mathbf{z} + \mathbf{b}^T\mathbf{S}^{-1}\mathbf{z} + c = \mathbf{z}^T\mathbf{z} + \mathbf{b}^T\mathbf{S}^{-1}\mathbf{z} + c,
$$

which is simply a sum of squares plus a constant (after we complete the squares). Thus the level surfaces in the new coordinates are indeed concentric spheres, and n orthogonal line searches in the new variables will arrive at the minimum point.

If the vectors $\mathbf{z}_1$ and $\mathbf{z}_2$ are orthogonal, their dot product $\mathbf{z}_1{}^T \mathbf{z}_2$ is zero. How can we calculate this dot product, in the original coordinate system? It's surprisingly easy:

$$\mathbf{z}_1^T \mathbf{z}_2 = (\mathbf{Sx}_1)^T (\mathbf{Sx}_2) = \mathbf{x}_1^T \mathbf{S}^T \mathbf{Sx}_2 = \mathbf{x}_1^T \mathbf{Ax}_2. \tag{4.7}$$

The calculation just involves $\mathbf{A}$, and we can dispense with the ghastly eigenvalues and diagonalizations that led to its derivation!

$$\mathbf{x}_1 \text{ is conjugate to } \mathbf{x}_2 \text{ if } \mathbf{x}_1^T \mathbf{Ax}_2 = 0.$$

Recall that the Gram–Schmidt algorithm tells us how to modify a "new" vector, $\mathbf{z}_{n+1}$, so that the result is orthogonal to each of a set of "old" vectors $\mathbf{z}_1, \mathbf{z}_2, \cdots, \mathbf{z}_n$:

If $\mathbf{z}_1, \mathbf{z}_2, \cdots, \mathbf{z}_n$ are mutually orthogonal vectors, the vector

$$\mathbf{z}_{n+1} - \frac{\mathbf{z}_{n+1}^T \mathbf{z}_1}{\mathbf{z}_1^T z_1} \mathbf{z}_1 - \frac{\mathbf{z}_{n+1}^T \mathbf{z}_2}{\mathbf{z}_2^T z_2} \mathbf{z}_2 - \cdots - \frac{\mathbf{z}_{n+1}^T \mathbf{z}_n}{\mathbf{z}_n^T \mathbf{z}_n} \mathbf{z}_n$$

will either be zero or orthogonal to each of $\mathbf{z}_1, \mathbf{z}_2, \cdots, \mathbf{z}_n$.

Replacing dot products as directed in (4.7), then, we have the rule for constructing conjugate directions.

If $\mathbf{x}_1, \mathbf{x}_2, \cdots, \mathbf{x}_n$ are mutually conjugate vectors, the vector

$$\mathbf{x}_{n+1} - \frac{\mathbf{x}_{n+1}^T \mathbf{Ax}_1}{\mathbf{x}_1^T \mathbf{Ax}_1} \mathbf{x}_1 - \frac{\mathbf{x}_{n+1}^T \mathbf{Ax}_2}{\mathbf{x}_2^T \mathbf{Ax}_2} \mathbf{x}_2 - \cdots - \frac{\mathbf{x}_{n+1}^T \mathbf{Ax}_n}{\mathbf{x}_n^T \mathbf{Ax}_n} \mathbf{x}_n$$

will either be zero (in which case we simply drop it) or conjugate to each of $\mathbf{x}_1, \mathbf{x}_2, \cdots, \mathbf{x}_n$.

Thus implementing (4.7) we have the conjugate gradient algorithm, replacing the gradient by a vector that is conjugate to all preceding search directions.

References

Davis, H.F., Snider, A.D.: Introduction to Vector Analysis, 7th edn. Hawkes Learning (1995)

Gecan, A., Snider, A.D.: Rationale for the Camerini-Fratta-Maffioli modified subgradient deflection and its ABS formulation. In: Proceedings of 2nd International Conference on ABS Algorithms, Beijing, pp. 57–66, June 1995 (1995)

Golub, G.H., O'Leary, D.P.: Some history of the conjugate gradient and Lanczos algorithms: 1948–1976. SIAM Rev. **31**(1), 50–102 (1989)

Lam, A.: BFGS in a Nutshell: An Introduction to Quasi-Newton Methods. https://towardsdatascie nce.com/bfgs-in-a-nutshell-an-introduction-to-quasi-newton-methods-21b0e13ee504. Accessed 2 May 2023

Constrained Optimization

5

5.1   The Karush–Kuhn–Tucker–John Conditions

To introduce the fundamental principle of constrained optimization we'll start with a simple physical situation, and increase the complexity by adding constraints.

Suppose your nose is cold, and the warmest point in the room is near the center, about 4 feet above the ground. So you walk there and bend over to get your nose warm fastest. What can we say about the temperature gradient at this, the warmest point, in the room?

If the gradient were not zero, it would point in a direction of steepest temperature ascent. But at the maximum, there is no direction in which you can move to increase the temperature. Therefore the gradient is zero at this maximum point.

We knew that. In Fig. 5.1 the dashed curves represent isotherms (level curves for the temperature function), the gradient directions are orthogonal to them, and at the maximum there is no direction for the gradient to point.

But suppose the warmest point is 6 feet above the floor, and you can only reach 5 feet high with your nose. *Now* what can we say about the gradient, at the warmest point you can reach?

Figure 5.2 demonstrates the situation. The blue line is the boundary of the constraint region (height < 5 feet). The woman in the purple skirt is not at the optimal point, because she can slide to the right and warm her nose more. But for you (the woman in blue), *the temperature gradient has no component along the constraint boundary tangent plane*—it is orthogonal to the boundary, and hence parallel to the constraint normal:

$$\text{temperature gradient} = \lambda \times (\text{normal to height constraint})$$

Your nose is at the optimal point, subject to the constraint.

© The Author(s), under exclusive license to Springer Nature Switzerland AG 2023
A. D. Snider, *Basics of Optimization Theory*, Synthesis Lectures on Mathematics
& Statistics, https://doi.org/10.1007/978-3-031-29219-4_5

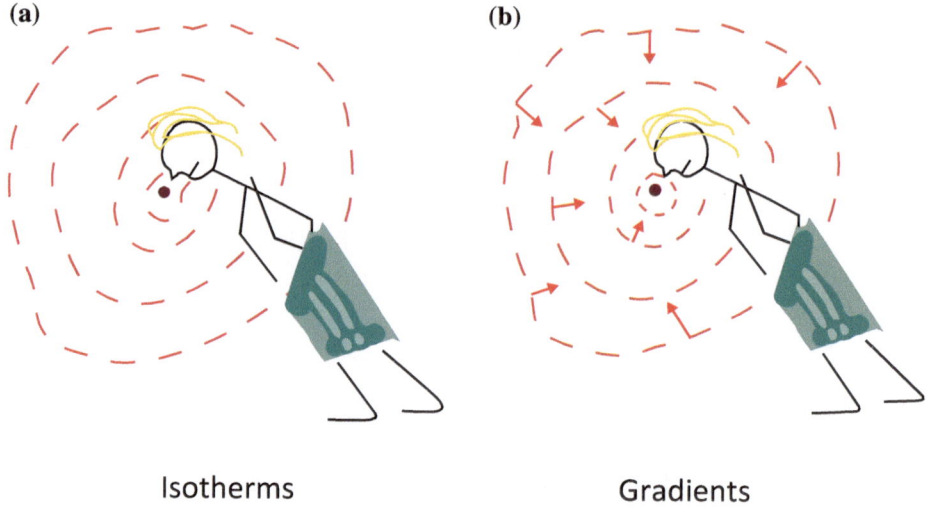

Fig. 5.1 There is no gradient direction at an unconstrained maximum

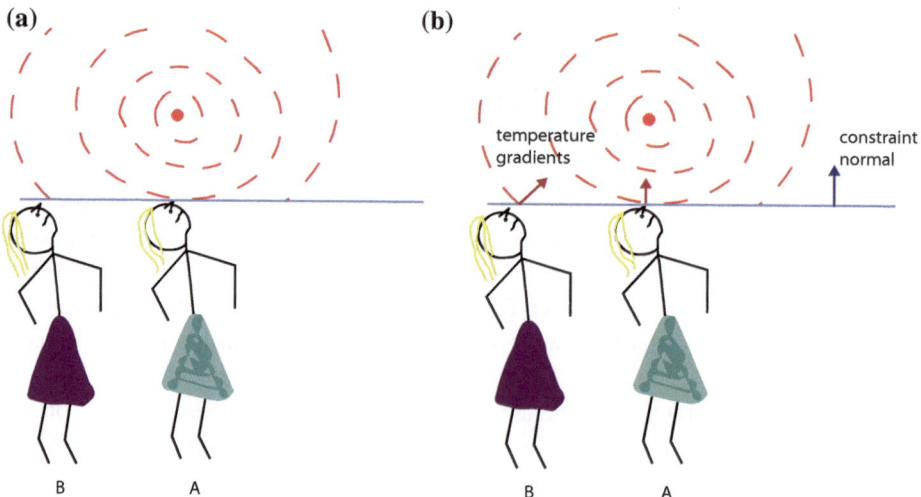

Fig. 5.2 Characteristics of a constrained maximum

OK, let's add another constraint. Your husband is painting the floor, and he has just finished the portion shown in Fig. 5.3.

You (the wife) are depicted situated in the optimal position; the temperature gradient is not zero, but you can't take advantage of it; you can't move vertically because of the height constraint and you can't move to the right because of the wet-paint constraint.

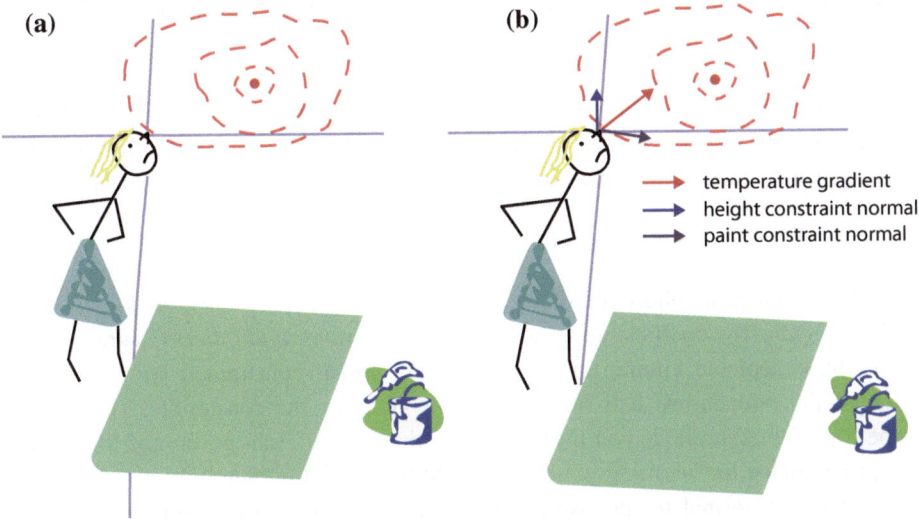

Fig. 5.3 Two constraints

How can we characterize the relation between the temperature gradient and the two constraint normals? Note that the constraints allow you to move your nose along a line perpendicular to the two constraint normals (perpendicular to the page). Therefore if the temperature gradient had a component along this line, you could warm your nose further by moving in that direction. So at the *optimal* point, the temperature gradient has no component perpendicular to the normals. In matrix parlance, it is *spanned* by these normals.

$$\text{temperature gradient} = \lambda_1 \times (\text{normal to height constraint})$$
$$+ \lambda_2 \times (\text{normal to wet-paint constraint}) \tag{5.1}$$

This is the *Karush–Kuhn–Tucker–John necessary condition* for a constrained maximum, in this situation. Note that we have chosen the normals for each constraint to point in the directions of constraint *violations*, and this makes the scalars λ_1 and λ_2 nonnegative.

Now if we express the height constraint as $g_1(x, y, z) \leq 0$ (with $g_1(x, y, z) = z - 5$ for the present circumstances), then a suitable normal to the constraint boundary (where $g_1(x, y, z) = 0$) is given by the gradient $\nabla g_1(x, y, z)$; after all, since the constraint was expressed $g_1 \leq 0$ and ∇g_1 points in the steepest *ascent* direction, the gradient of g_1 points in the direction of constraint violation. Treating the wet-paint constraint similarly ($g_2(x, y, z) \leq 0$) and using $f(x, y, z)$ to denote the objective function (temperature), we rewrite (5.1) in the convenient form

$$\nabla f = \lambda_1 \nabla g_1 + \lambda_2 \nabla g_2. \tag{5.2}$$

Equation (5.2) is our basic tool for analyzing constrained optimization problems. To make it more elegant, let's list a few observations.

(i)   Although the constrained optimization conditions were not formulated in full generality until 1939 (Karush 1939; John 1948; Kuhn and Tucker 1951), the topic has a long history. Indeed, (5.2) was expressed by Lagrange and his colleagues in the form

$$\nabla(f - \lambda_1 g_1 - \lambda_2 g_2) = 0, \tag{5.3}$$

and the function whose gradient vanishes, $f - \lambda_1 g_1 - \lambda_2 g_2$, is known as the *Lagrangian.* The coefficients λ_1, λ_2 are called *Lagrange multipliers* (Lagrange 1806).

(ii) We have depicted circumstances where the optimal constrained point lies on the constraint boundaries; it is on the verge of violating the constraints. This need not be true; for example, if you husband had painted a different portion of the floor, the optimal place for warming your nose could look like Fig. 5.4.

    Now the normal to the wet-paint constraint plays no role, and the optimum is the same as in Fig. 5.2. The term $\lambda_2 g_2$ is not needed in Eq. (5.3). The following nomenclature simply extends a notion we introduced in Sect. 2.6:

A constraint $g(x, y, z) \leq 0$ is said to be *active* at the points where $g(x, y, z) = 0$, and *inactive* at points where $g \leq 0$.

Fig. 5.4 Inactive constraint

Fig. 5.5 Active constraint
with zero Lagrange multiplier

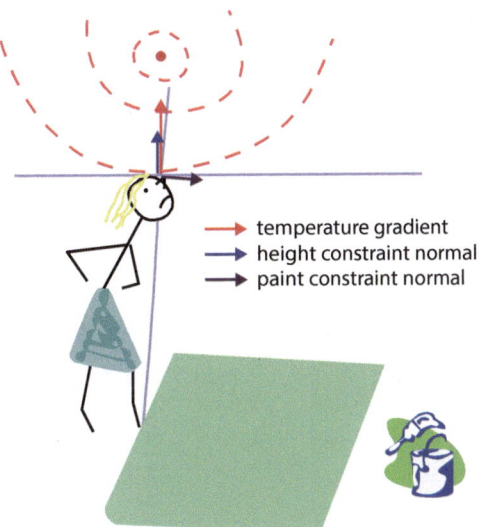

temperature gradient
height constraint normal
paint constraint normal

A slick way of formulating the constrained optimization condition is to *include* a term λg in the Lagrangian for *every* constraint, but to note that the Lagrange multiplier λ is zero for the constraints that are inactive at the optimum point, while $\lambda \geq 0$ for the active constraints.

(iii) The sign rule $\lambda \geq 0$ may strike you as a little conservative; surely the Lagrange multipliers called for by Fig. 5.3 are positive—strictly greater than zero. However there are some exceptional circumstances that force a multiplier to be zero even when its constraint is active. Suppose, for example, your husband had painted the floor right up to the place where you would have been standing if there were no wet-paint constraint; see Fig. 5.5.

   Then the wet-paint constraint is active, but the temperature gradient is parallel to the height-constraint normal and there is no need for the paint-constraint normal in Eq. (5.1); $\lambda_2 = 0$.

(iv) Another sign-rule anomaly occurs if there is a linear relationship among the active constraint normals at the optimum point, i.e. they are linearly dependent.

$$a_1 \nabla g_1 + a_2 \nabla g_2 + a_3 \nabla g_3 = 0 \quad \text{(and not all } a_i = 0\text{)}$$

For example suppose there is a hole in your floor, and the inequality $g_3 \leq 0$ expresses the condition that you can't stand in the hole. Suppose further your husband has just finished painting around the hole (Fig. 5.6).

   This coincidence would make the wet-paint-constraint normal, ∇g_2, parallel to the hole-constraint normal, ∇g_3; say, $2\nabla g_2 - \nabla g_3 = 0$. Then although it is still true that the temperature gradient ∇f can be expressed as a linear combination of the (active) height-constraint normal ∇g_1 and the (active) wet-paint-constraint normal

Fig. 5.6 Linearly dependent active constraint normals

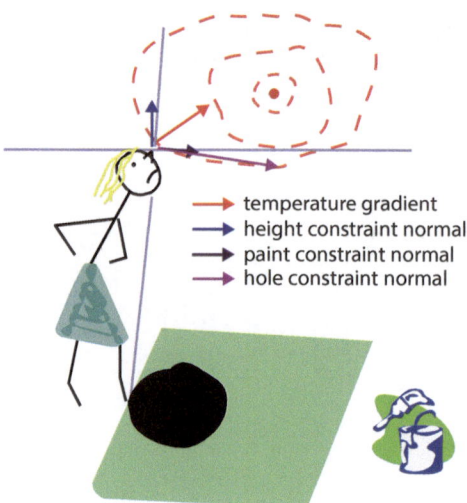

temperature gradient
height constraint normal
paint constraint normal
hole constraint normal

∇g_2 with *positive* Lagrange multipliers

$$\nabla f = \lambda_1 \nabla g_1 + \lambda_2 \nabla g_2, \quad (5.2 \text{ repeated})$$

it is also true that we can add zero (disguised as $(2\nabla g_2 - \nabla g_3)$) to this and get the temperature gradient re-expressed in terms of active-constraint normals with Lagrange multipliers *of both signs*:

$$\nabla f = \lambda_1 \nabla g_1 + (\lambda_2 + 2)\nabla g_2 - \nabla g_3.$$

Therefore the sign-rule claim that each $\lambda_i \geq 0$ in the expression $\nabla f = \lambda_1 \nabla g_1 + \lambda_2 \nabla g_2 + \cdots$ can only be assured if all the active-constraint normals ∇g_i are linearly independent.

(v) A constraint of the form $g(x, y, z) \geq 0$ can, of course, be converted to the less-than-or-equal form by multiplying by -1. But what about *equality* constraints, $h(x, y, z) = 0$? It makes no sense to speak of the "normal pointing in the direction of constraint violation" since *both* normal directions ($\pm$) violate this constraint. (Indeed, $h(x, y, z) = 0$ is equivalent to *two* inequality constraints $h(x, y, z) \leq 0$ and $-h(x, y, z) \leq 0$ with oppositely-directed constraint-violating normals.) So for equality constraints $h(x, y, z) = 0$ we include a term μh in the criterion (5.2), but we can not impose any conditions on the sign of the Lagrange multiplier μ.

So here (at last) is a correct statement of the Karush–Kuhn–Tucker–John necessary conditions for a maximum point:

Let $\mathbf{x}$ be a point in N-dimensional space, $\mathbf{x} = (x_1, x_2, \cdots, x_N)$, and let $f, g_1, g_2, \cdots, g_M, h_1, h_2, \cdots, h_P$ be continuously differentiable functions of $\mathbf{x}$ in the region constrained by

$$g_1(x) \le 0, g_2(x) \le 0, \ldots, g_M(x) \le 0. \tag{5.4}$$

Assume $\mathbf{x}^*$ is the maximum point of $f(\mathbf{x})$ among all $\mathbf{x}$ satisfying the inequality constraints (5.4) *and* the equality constraints

$$h_1(\mathbf{x}) = 0, h_2(\mathbf{x}) = 0, \ldots, h_P(\mathbf{x}) = 0, \tag{5.5}$$

and also assume that the set of gradients $\{\nabla h_i(\mathbf{x}^*), i = 1, 2, \ldots, P\}$ and the subset of gradients $\{\nabla g_i(\mathbf{x}^*), i = 1, 2, \ldots, M\}$ such that $g_i(\mathbf{x}^*) = 0$ are linearly independent.

Then there exist constants (Lagrange multipliers) $\lambda_1, \lambda_2, \ldots, \lambda_M, \mu_1, \mu_2, \ldots, \mu_P$ such that the following $(N + M + P)$ equations hold at $\mathbf{x}^*$:

$$\nabla\big(f - \lambda_1 g_1 - \lambda_2 g_2 - \cdots - \lambda_M g_M - \mu_1 h_1 - \mu_2 h_2 - \cdots - \mu_p h_P\big) = \mathbf{0}$$
$$(N \text{ equations}) \tag{5.6}$$

$$\lambda_1 g_1 = 0, \lambda_2 g_2 = 0, \cdots, \lambda_M g_M = 0 \quad (\lambda_i = 0 \text{ or } g_i = 0 \text{ or both})$$
$$(M \text{ equations}) \tag{5.7}$$

$$h_1 = 0, h_2 = 0, \ldots, h_P = 0 \quad (\text{Eq. (5.5) at } \mathbf{x}^*)$$
$$(P \text{ equations}). \tag{5.8}$$

Moreover if $g_i(\mathbf{x}^*) = 0$ then $\lambda_i \ge 0$.

Equations (5.6)–(5.8) can be regarded as $(N + M + P)$ simultaneous equations in $(N + M + P)$ unknowns $\big(x_1^*, x_2^*, \ldots, x_N^*, \lambda_1, \lambda_2, \ldots, \lambda_M, \mu_1, \mu_2, \ldots, \mu_P\big)$. Typically they will have a finite number of solutions, one of which will be the maximum point. We can immediately eliminate the solutions for which any $g_i(\mathbf{x}^*)$ is positive or any λ_i is negative, and compare the values of $f(\mathbf{x}^*)$ among those remaining.

Do we have an algorithm to solve these equations? Yes; observe that if we set the "complete" gradient (*including the partials with respect to the unknowns* λ_i, μ_j) of the *Lagrangian*

$$F(x, \lambda, \mu) \equiv f - \lambda_1 g_1 - \lambda_2 g_2 - \cdots - \lambda_M g_M - \mu_1 h_1 - \mu_2 h_2 - \cdots - \mu_p h_P$$

equal to zero, we reproduce Eqs. (5.6)–(5.8) almost exactly:

$$\frac{\partial F(x, \lambda, \mu)}{\partial x_i} = 0, i = 1, 2, \ldots, N \quad (\text{Eq. (5.6)})$$

$$\frac{\partial F(x, \lambda, \mu)}{\partial \mu_i} = h_i(x) = 0, i = 1, 2, \ldots, P \quad (\text{Eq. (5.8)})$$

$$\frac{\partial F(x, \lambda, \mu)}{\partial \lambda_i} = g_i(x) = 0, i = 1, 2, \ldots, M \quad \text{(Eq. (5.7) when } \lambda_i \neq 0).$$

(And if $\lambda_i = 0$ the equation for g_i is moot.)

But don't forget we just spent a whole chapter on how to solve a system of equations that expresses "gradient equals zero"! So we have reduced the *constrained* optimization problem to an *unconstrained* optimization problem (with more unknowns).

The general approach to solving constrained optimization problems, then, is to construct the Lagrangian and to add, to the "unknowns" list, the Lagrange multipliers λ_i for the inequality constraints and μ_i for the equality constraints. We arbitrarily select a subset of the *inequality* constraints and set the corresponding $\lambda_i = 0$, drop the g_i from the Lagrangian, and use the unconstrained techniques (steepest descent, conjugate gradient, etc.) to find the values of $\mathbf{x}$, λ, and μ where the (extended) gradient of the Lagrangian equals zero. Next we check the conditions $g_i(\mathbf{x}) \leq 0, \lambda_i \geq 0$ for this solution; if they are violated, we disqualify this solution.

Then we select *another* subset of the inequalities to eliminate from the Lagrangian and solve again (for zero gradient). And so on. If the constrained problem has an optimal point, it must be one of the retained solutions.

Specialists have formulated guidelines for efficiently sorting through the subsets of the inequalities. We refer the reader to the advanced literature.

5.2   Examples

Our first example is quite elementary. It can be worked by the basic calculus method described in Sect. 3.1. However we approach it through the Karush–Kuhn–Tucker–John logic, to gain some insight into the latter.

Example 5.1 Maximize $(x - 1)^3$ subject to $x \leq 2$ and $x \geq 0$ (Fig. 5.7).

Solution. The derivative, $3(x - 1)^2$, is zero at the interior inflection point, so this is an end-point maximum situation. The maximum, 1, obviously occurs at $x = 2$.

The KKTJ logic starts with writing the Lagrangian, with the inequality $x \geq 0$ properly expressed as $-x \leq 0$:

$$L = (x - 1)^3 - \mu_1(x - 2) - \mu_2(-x).$$

The gradient of the Lagrangian is zero:

$$\nabla L = \partial L / \partial x = 3(x - 1)^2 - \mu_1 + \mu_2.$$

Case 1. Both inequalities active: $x = 2$ and $x = 0$. Impossible.

Fig. 5.7 Example 5.1

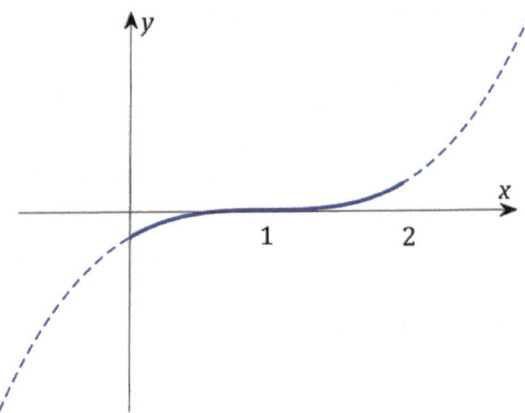

Case 2. The first constraint is active: x equals 2 and μ_2 is zero.

$$3(2-1)^2 - \mu_1 + (0) = 0, \text{ so } \mu_1 = 3.$$

The Lagrange multiplier is nonnegative, so $x = 2$ is a viable candidate.
Case 3. The second constraint is active: x equals 0 and μ_1 is zero.

$$3(0-1)^2 - (0) + \mu_2 = 0, \text{ so } \mu_2 = -3.$$

The negative Lagrange multiplier disqualifies this candidate.
Case 4. Both constraints are inactive: $\mu_1 = \mu_2 = 0$ and $3(x-1)^2 = 0$, so $x = 1$. All is well with KKTJ and this candidate is also viable.

Of the two viable candidates, $x = 1$ and $x = 2$, the latter yields the higher value for $(x-1)^3$; the maximum does indeed occur at the end point.

Example 5.2 Maximize $4x_1 - x_1^2 + 6x_2 - x_2^2 + (x_1 + x_2)^2 + x_3$ subject to

$$x_1 + x_2 \le 3,$$

$$-2x_1 + x_2 \le 2,$$

$$x_1^2 + x_2^2 + 2x_1 x_2 + x_3 = 0.$$

Solution. The Lagrangian is

$$L = 4x_1 - x_1^2 + 6x_2 - x_2^2 + (x_1 + x_2)^2 + x_3 - \mu_1(x_1 + x_2 - 3)$$
$$- \mu_2(-2x_1 + x_2 - 2) - \lambda_1\left(x_1^2 + x_2^2 + 2x_1 x_2 + x_3\right).$$

The gradient condition becomes

$$0 = \frac{\partial L}{\partial x_1} = 4 - 2x_1 + 2(x_1 + x_2) - \mu_1 + 2\mu_2 - \lambda_1(2x_1 + 2x_2)$$

$$0 = \frac{\partial L}{\partial x_2} = 6 - 2x_2 + 2(x_1 + x_2) - \mu_1 - \mu_2 - \lambda_1(2x_2 + 2x_1)$$

$$0 = \frac{\partial L}{\partial x_3} = 1 - \lambda_1$$

Case 1. All constraints active:

$$x_1 + x_2 = 3,$$

$$-2x_1 + x_2 = 2,$$

$$x_1^2 + x_2^2 + 2x_1 x_2 + x_3 = 0$$

The solution to these 6 equations is

$$x_1 = \frac{1}{3}, \ x_2 = \frac{8}{3}, \ x_3 = -9, \ \mu_1 = \frac{14}{9}, \ \mu_2 = -\frac{8}{9}, \ \lambda_1 = 1.$$

The negative Lagrange multiplier μ_2 disqualifies this candidate.

Case 2. The first inequality constraint is active.

$$x_1 + x_2 = 3,$$

$$\mu_2 = 0,$$

$$x_1^2 + x_2^2 + 2x_1 x_2 + x_3 = 0.$$

The solution to the equations is

$$x_1 = 1, \ x_2 = 2, \ x_3 = -9, \ \mu_1 = 2, \ \mu_2 = 0, \ \lambda_1 = 1.$$

This solution qualifies as a candidate.

Case 3. The second inequality constraint is active.

$$-2x_1 + x_2 = 2,$$

$$\mu_1 = 0$$

$$x_1^2 + x_2^2 + 2x_1x_2 + x_3 = 0.$$

The solution to the equations is

$$x_1 = 0.8, \ x_2 = 3.6, \ x_3 = -19.36, \ \mu_1 = 0, \ \mu_2 = -1.2, \ \lambda_1 = 1.$$

The negative Lagrange multiplier μ_2 disqualifies this candidate.

Case 4. Neither inequality constraint is active.

$$\mu_1 = \mu_2 = 0$$

$$x_1^2 + x_2^2 + 2x_1x_2 + x_3 = 0.$$

The solution to these equations is

$$x_1 = 2, \ x_2 = 3, \ x_3 = -25, \ \mu_1 = 0, \ \mu_2 = 0, \ \lambda_1 = 1.$$

These values of x_1 and x_2 violate the first inequality constraint, disqualifying this candidate.

The only candidate that meets the KKTJ conditions is $x_1 = 1, x_2 = 2, x_3 = -9$. Therefore the maximum value of the objective function equals

$$4(1) - (1)^2 + 6(2) - 2^2 + (1+2)^2 + (-9) = 11.$$

References

John, F.: Extremum problems with inequalities as side conditions. In: Friedrichs, K.O., Neugebauer, O.E., Stoker, J.J. (eds.) Studies and Essays, Courant Anniversary Volume. Wiley-Interscience (1948)

Karush, W.: Minima of functions of several variables with inequalities as side conditions. Master's thesis, Department of Mathematics, University of Chicago (1939)

Kuhn, H.W., Tucker, A.W.: Nonlinear programming In: Neyman, J. (ed.) Proceedings of Second Berkeley Symposium on Mathematical Statistics and Probability. University of California Press (1951)

Lagrange, J. L.: Leçons sur le calcul des fonctions, 2nd ed. Reprinted as, Oeuvres, vol. 9 (1806)

What's Left?

6

Now you know the basics of optimization theory. Of course a subject this important has been explored and fathomed in many directions—some theoretical, some prompted by practical considerations. Here is a brief listing of the issues you may want to confront, to become more expert in the discipline.

1. *Other nomenclature and notation.* We have embraced two- and three-dimensional examples to facilitate *visualizing* the various concepts arising in optimization theory; trying to express its formulas in full generality can distract a novice from grasping the basics. As you progress to further readings you will quickly get the knack of shifting between abstract and specific formulations. There are some nomenclature variations to watch for: our diamond model for the search region in a linear program generalizes to a "convex polyhedron" in n dimensions, bounded by "supporting hyperplanes". Our "legitimate corners" are often called "feasible" points of intersection. Some authors refer to the Simplex tableau as a "dictionary".

2. *Foundations.* We have deliberately been casual about the degree of rigor employed in our exposition, because we believe scrutinizing each statement for minimal conditions and maximal generality often introduces stumbling blocks to the initial comprehension of novel ideas; it can also put a damper on the excitement of discovering new mathematical toys. Be assured that the literature is replete with precise justifications of the properties of convexity that we have taken for granted, the limitations of neglecting higher-order terms in approximating nonlinear objective functions, and the like.

3. *Efficiency.* Although we have discussed the major steps taken to upgrade the efficiency of the Simplex algorithm, there is room for improvement. Take, for example, the persistence of all those identity matrix columns in each tableau; sparse matrix coding can reduce this wasted real estate. The s/t competition invites some look-ahead strategies that can speed the solution. Well-known principles of numerical analysis can be

A. D. Snider, *Basics of Optimization Theory*, Synthesis Lectures on Mathematics & Statistics, https://doi.org/10.1007/978-3-031-29219-4_6

brought to bear, to trim the rounding errors accumulated through extensive pivoting and to estimate the sensitivity of the solution to perturbations in the data. And certain types of applications such as transportation and network problems can be tailored to conform to the linear programming scheme efficiently. Sometimes, however, the improvements proffered by these strategies are not worth the effort it takes to implement them.

4. *Degeneracy.* Research has focused on: (*a*) given a tiebreaker rule, find a tableau that will exhibit cycling when that rule is applied; (*b*) find tiebreaker rules that always prevent cycling; (*c*) considering the anomalous nature of degeneracy, decide when tiebreaker rules are worthwhile.

5. *Primal-Dual algorithms.* These exploit the mutual bounds provided by the dual linear program (Sect. 2.12) to devise improved Simplex strategies that switch between the primal and dual problems.

6. *Convergence.* Readers with industrial experience have probably noted that our citations of convergence of the algorithms share a little of the naivete of the traditional epsilon-delta proclamations of classical mathematics. The Simplex algorithm is guaranteed to locate the maximum objective in a *finite* number of steps (Sect. 2.1); each iteration of Newton's method or the method of false position ultimately reduces a *multiple* of the distance to the optimum point by the power "*s*" (Sect. 3.2); it has been proved that worst-case linear programs with *n* variables can force Simplex to take {more than any power of *n*} pivots to find the maximum (Klee and Minty 1972). These academically profound statements don't directly address the issue of how close we can get in practice, with limited funding. Some rules of thumb have been established. "Interior point" algorithms, that do not confine themselves to the edges of the diamond, have received a lot of attention.

7. *Multiple local maxima and minima.* The higher-dimensional expositions in Chap. 4 are focused on speedily converging to an objective function's minimum point from an "initial guess" that is close enough to the minimum for the objective to be approx-imated as a quadratic function. But the heartbreaking reality is that many important applications possess objective functions with *multiple* local maxima and minima, and once a candidate point falls within the "basin of attraction" of any of these, our dedi-cated algorithms simply drive it closer—unaware that the *global* minimum may reside somewhere else. Much research has focused on robust strategies, such as inserting occasional "breather" steps that crudely sample nearby regions for lower objective values—necessitating a shift and a fresh start for the iterations.

8. *Constraints.* Instead of directly confronting constraints via the Karush–Kuhn–Tucker–John equations (Sect. 5.1), one can artificial add weighted *penalty* or *barrier* terms to an objective function that steer the *unconstrained* search algorithms of Chap. 4 away from regions that violate the constraints.[1]

---

[1] This logic is inspired by the practice of enforcing income tax deadlines by imposing extreme interest fees on late returns.

Optimization theory remains a fertile ground for research and implementation. We commend you to its study.

Reference

Klee, V., Minty, G.: How good is the simplex algorithm? In: Shisha, O. (ed.) Inequalities III, pp. 159–175. Academic Press, NY (1972)